Antimicrobial Stewardship - New Insights

Edited by Ghulam Mustafa

Published in London, United Kingdom

Antimicrobial Stewardship - New Insights
http://dx.doi.org/10.5772/intechopen.111192
Edited by Ghulam Mustafa

Contributors
Abigarl Ndudzo, Agrima Bhatt, Anusha Gopinath, Duduzile Ndhlovu, Esma Rabvukwa, Faithful Miebaka Daniel, Fatema Rampurwala, Fatima Rehan, Ghulam Mustafa, Hadi Sajid Abdulabbas, Hafiza Salaha Mahrosh, Huda Najh Al-Baroody, Hussein Al-Dahmoshi, Ibinabo Laura Oboro, Kakithakara Vajravelu Leela, Khalifa Musa Muhammad, Kolapo Fasina, Maheswary Datchanamoorthy, Manasi Mishra, Mansurat Oluwatoyin O. Shoge, Maryam Nasrumminallah, Nachiket Garge, Narayanasamy Damodharan, Nesisa Nyathi, Nishat Sulemani, Noor Salman Al-Khafaji, Olanrewaju O. Popoola, Pankti Dhumal, Radhakrishnan Rahul, Rehan Deshmukh, Rini S. Handayani, Ruogu Tang, Sadia Khan, Shikha Gaikwad, Simeon Chijioke Amadi, Sokeipirim Ikiroma, Srashti Bajpai, Suhad Y. Abed, Temitope Oyedemi, Tolulope Fadeyi, Utpal Roy, Vita Pertiwi

First published in London, United Kingdom, 2024 by IntechOpen
IntechOpen is the global imprint of INTECHOPEN LIMITED, registered in England and Wales, registration number: 11086078, 167-169 Great Portland Street, London, W1W 5PF, United Kingdom

British Library Cataloguing-in-Publication Data
A catalogue record for this book is available from the British Library

Additional hard and PDF copies can be obtained from orders@intechopen.com

Antimicrobial Stewardship - New Insights
Edited by Ghulam Mustafa
p. cm.

This title is part of the Pharmaceutical Science Book Series, Volume 4
Topic: Pharmacology
Series Editor: Rosario Pignatello
Topic Editor: Cristina Manuela Drăgoi
Associate Topic Editor: Alina Crenguţa Nicolae

Print ISBN 978-1-83769-090-9
Online ISBN 978-1-83769-091-6
eBook (PDF) ISBN 978-1-83769-092-3
ISSN 3033-3318

For EU product safety concerns:
IN TECH d.o.o., Prolaz Marije Krucifikse Kozulić 3, 51000 Rijeka, Croatia,
info@intechopen.com or visit our website at intechopen.com.

IntechOpen Book Series

Pharmaceutical Science

Volume 4

Aims and Scope of the Series

Pharmaceutical science focuses on the design, synthesis, formulation, targeting, distribution, safety, and efficacy of active compounds as potential therapeutics. It is a large interdisciplinary discipline that aims to integrate the basic principles of physical and organic chemistry, biochemistry, biology, and engineering to discover, develop, and characterize active compounds and to optimize the formulation and delivery of drugs in the body for offering new and improved safe and efficacious therapies against human diseases. The research areas covered by the pharmaceutical sciences range from medicinal chemistry and pharmaceutical technology to pharmacology and toxicology, which represent the preliminary phases of drug development. Medicinal chemistry involves the design and synthesis of pharmaceuticals as well as the isolation of active agents from natural sources. Computer-aided strategies are increasingly involved in this drug discovery process. Pharmaceutics is a multidisciplinary science that examines the relationships between drug formulation, delivery, distribution, and clinical outcomes. Modern clinical approaches are increasingly relying on controlled release strategies and drug delivery and targeting systems, including nanotechnological platforms (nanomedicine). Pharmacology is the science of drug action in biological systems. Pharmacologists also make drugs as tools to explore aspects of cell and tissue functions. Toxicology is the study of the adverse effects of active agents on living organisms and the ecosystem, including the prevention and amelioration of such adverse effects. This book series includes volumes on Drug Discovery, Delivery, and Pharmacology. Their overall aim is to present the latest research in the whole path of drug discovery and development from different points of view of this multidisciplinary and dynamic field.

Meet the Series Editor

Prof. Rosario Pignatello is a Full Professor of Pharmaceutical Technology and Legislation at the University of Catania, Italy. He is the Director of the Department of Drug and Health Sciences. He has nearly 30 years of experience in the research and development of innovative formulations for the controlled release and targeting of bioactive molecules, through chemical approaches as well as nanotechnological carriers, aimed at treating different disorders. Prof. Pignatello has coauthored about 180 papers and edited a series of textbooks on biomaterials and their application in medicine. The main areas of his research are polymeric and lipid-based micro- and nanoparticles as modified drug delivery systems; vesicular nanocarriers (liposomes, micelles); lipophilic prodrugs and conjugates; synthesis and evaluation of new polymeric biomaterials for drug delivery and tissue regeneration. In particular, Prof. Pignatello works actively in the field of ocular drug delivery, leading the Research Centre for Ocular Nanotechnology, within the NANOMED Centre (Centre for Nanomedicine and Pharmaceutical Nanotechnology) at the University of Catania.

Meet the Volume Editor

Dr. Ghulam Mustafa obtained his Ph.D. in Biochemistry from the University of Agriculture Faisalabad, Pakistan, with research work from the University of California, San Diego, USA. He has a strong biochemistry background with extensive research and technology development experience in biochemistry, microbiology, biotechnology, and bioinformatics. He has been working on drug discovery using both traditional and in silico approaches. Dr. Mustafa has contributed many publications to journals of international repute. He is an academic editor for many international journals and their special issues. He has participated in numerous conferences, seminars, and congresses and has authored many books and book chapters. He has also delivered hands-on training in several workshops across the country in the areas of microbiology, molecular biology, and bioinformatics. He is a fellow of several international societies. Currently, he is an Associate Professor of Biochemistry at Government College University Faisalabad, Pakistan. He has more than 13 years of experience in teaching and research.

Contents

Preface

This book, *Antimicrobial Stewardship – New Insights*, offers valuable insights into antibiotics for researchers, academicians, pharmacists, physicians, nurses, healthcare workers, technicians, and other interested readers. In the early days of antibiotics, even as antibiotic resistance developed, a new antibiotic was always available for the treatment of increasingly resistant bacterial species. A considerable decrease in the development of new antibiotics has been observed during the past 40 years and, therefore, the options to treat increasingly resistant infections are now limited. Infections caused by antibiotic-resistant pathogens are causing millions of deaths worldwide. Antimicrobial stewardship refers to a series or program of interventions for monitoring the direct use of antimicrobials at healthcare institutions. Antimicrobial stewardship is of the utmost importance for optimizing antimicrobial uses to prevent or slow down antimicrobial resistance.

This book is organized into two sections: "Importance of Antibiotic Stewardship in Healthcare Settings" and "Antimicrobial Resistance and Antimicrobial Stewardship". In the first section, the introductory chapter provides a comprehensive overview of antimicrobial stewardship, which refers to interventions and control programs for the optimization of antibiotic usage. It also discusses the need for antimicrobial stewardship programs, strategies, evaluations, and international guidelines and approaches. Chapter 2 examines how antibiotic stewardship plays a role in the management of antibiotic resistance. The most important things include developing policies and regulations to support antibiotic stewardship in primary health care, optimizing available facilities and infrastructure, improving facilities to support antibiotic stewardship, increasing capable human resources, monitoring and evaluating antibiotic prescribing, and building system information on antibiotic resistance, which should be easily accessible to health workers in primary healthcare facilities. Chapter 3 explores the applications of antimicrobial stewardship and natural product chemistry to tackle antimicrobial resistance. It discusses how plants have contributed immensely to our traditional medicines as well as to the field of drug discovery due to the presence of plant bioactive secondary metabolites. The focus of Chapter 4 is the implementation of antimicrobial stewardship in various healthcare settings to reduce antimicrobial resistance. Chapter 5 discusses the role of antimicrobial stewardship in resource-limited settings, highlighting that the implementation of antimicrobial stewardship programs remains relatively limited and faces complex challenges in resource-poor healthcare settings. Chapter 6 shows how probiotics and their derivatives can play a role in the improvement of antimicrobial stewardship as it concerns human health. It investigates how the gut microbiota are modulated by probiotics, ghost probiotics, postbiotics, and synbiotics, and their importance in infection prevention and antibiotic stewardship. Finally, Chapter 7 is about laboratory information management system analysis, which is employed to support control of antimicrobial resistance.

The second section includes six chapters. Chapter 8 discusses the role of implementation of strict regulation and monitoring regarding the irrational use of antibiotics in animals, sewage disposal, waste disposal, and hospital infection control practices. The spread of antimicrobial resistance in the environment is also an alarming issue for the world. Therefore, Chapter 9 examines the role of antimicrobial stewardship in controlling antimicrobial resistance in the environment. Chapter 10 discusses how an effective antimicrobial stewardship is integral to managing various infections and preventing their further escalation through antimicrobial stewardship programs, de-escalation therapy, combination therapy, antibiotic dose optimization, and prophylactic antibiotic use in those who are at high risk of infection. Chapter 11 provides a short, general review of biocompatible polymer materials and their applications in antimicrobial fields. It begins with a background of microbial threats, discussing the demands of biocompatible polymers for antimicrobial applications, then moves on to describing the basic knowledge of biocompatible polymers, including definitions, advantages, and typical examples. The chapter ends with a discussion of some approaches to utilizing biocompatible polymers in antimicrobial applications. In Chapter 12, the authors discuss the inappropriate practices of prescribing antibiotics that have led to increased multidrug resistance in bacteria, envisaging the implementation of rigorous guidelines for proper use of antibiotics in human beings. Chapter 13 covers the patho-phylotyping of diarrheagenic *Escherichia coli* interaction along with antibiotic resistance and biofilm formation capacity.

Ghulam Mustafa
Department of Biochemistry,
Government College University Faisalabad,
Faisalabad, Pakistan

Section 1

Importance of Antibiotic Stewardship in Healthcare Settings

Chapter 1

Introductory Chapter: Antimicrobial Stewardship – Antibiotics Usage Optimization to Reduce the Microbial Resistance Burden

Hafiza Salaha Mahrosh and Ghulam Mustafa

1. Introduction to antimicrobial stewardship

Antimicrobial stewardship (AMS) refers to interventions and control programs for the optimization of antibiotic usage. Antimicrobial stewardship is growing fastly in numbers with diverse contexts ranging from hospitals to the veterinary community [1]. Despite much acceptance, the term is still facing many challenges in different communities. There are different ways to view antimicrobial stewardship such as a collection of coordinated interventions, a program, a philosophy, and an ethic [1]. Antimicrobial stewardship policies are mainly focused on the optimization of antibiotic usage to reduce antimicrobial resistance and improvement of patients' outcomes and hospital hygiene in different healthcare centers. This may involve creating criteria for the use of antibiotics, such as limiting their use to specific indications or restricting certain antibiotics to specific patient populations [2]. Various antimicrobial stewardship strategies have been mentioned in **Figure 1**.

Antimicrobial stewardship programs (ASPs) are typically led by multidisciplinary teams of healthcare professionals, including infectious diseases specialists, pharmacists, and microbiologists. These teams work together to develop and implement strategies for promoting the appropriate use of antimicrobial agents [3]. The primary strategies of ASPs include prospective auditing and feedback, pre-authorization, dose optimization, hospital education, computer monitoring, and antibiotic restriction. Regular evaluation of outcomes of these programs can help identify areas to further improve and ensure that the programs are meeting their desired goals [4].

Antimicrobial stewardship policies are important for promoting the responsible use of antibiotics and reducing the risk of antibiotic resistance. By implementing these policies, healthcare providers can help ensure that antibiotics remain effective for future generations. In this current article, the policies to cope with the ongoing problems with uprising challenges have been discussed for the understanding of the importance of antimicrobial stewardship in the medical sector for the management of antibiotic resistance.

IntechOpen

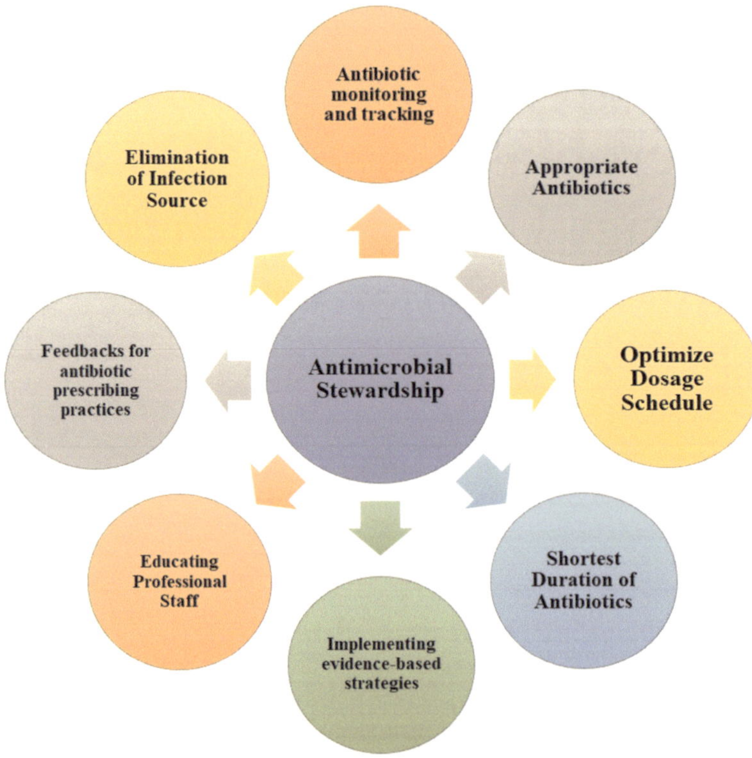

Figure 1.
Various antimicrobial stewardship strategies used to overcome antimicrobial resistance.

2. Antimicrobial resistance

Antimicrobial resistance is the ability of microorganisms to withstand the effects of antimicrobial drugs and it develops when an antibiotic loses its ability to effectively stop microbial growth. Antibiotic resistance is a growing global health concern, as it can lead to longer hospital stays, higher healthcare costs, and increased mortality rates. When antibiotics are overused or misused, bacteria are exposed to drugs unnecessarily, and those that survive develop resistance to the antibiotics used [5]. Antimicrobial resistance (AMR) jeopardizes not only public health but also economic growth and security. Based on high-level predictions and global reports, AMR has the potential to result in 10 million fatalities worldwide by 2050 [6]. The Centers for Disease Control and Prevention (CDC) has estimated that antibiotic resistance increases direct healthcare expenditures in the United States by $20 billion annually, excluding the estimated $35 billion in productivity losses [7].

The wide spread of antimicrobials has resulted in the expression of resistance to these antimicrobial agents. Resistance-encoding genes have been probably present for thousands of years, either as a defense against antibiotics or for unknown purposes, and the incorporation of these genes by human commensal and pathogenic flora has been quickly followed [8]. The mass production of antimicrobials has provided humanity with a temporary advantage in the battle against microorganisms however, if the current rate of increase in antimicrobial resistance continues, we may enter the postantibiotic era [8].

Various programs and strategies have been developed over the years to reduce antimicrobial resistance and promote the proper and effective use of antibiotics [9]. The antimicrobial stewardship programs (ASPs) have attracted a lot of attention as one of the most important strategies for combating AMR. Antimicrobial stewardship is a coordinated approach to promote the appropriate use of antibiotics and other antimicrobial agents in healthcare settings [3].

3. History

Since the 1940s, different educational programs, management strategies, clinical protocols, and guidelines have been developed by infectious-disease organizations to control and prevent microbial infections in various sectors [10]. In the United States, the Centers for Disease Control and Prevention (CDC) launched the first educational program in 2009 to advocate for the rational use of antibiotics in acute-care settings, and improved antibiotic use as a strategy was adopted in 2013 to address the problem [11]. In 2015, World Health Organization (WHO) issued a global action plan (GAP) on the appropriate use and effects of antibiotics on humans and animals along with the validity of antibiotics for future generations as the main aim of GAP [12].

AMS has a positive impact in hospitals with shorter lengths of stay, shorter treatment duration without an increase in mortality, and a decrease in colonization and infection with resistant bacteria. Many studies have been conducted on interventions aimed at outpatient prescribers proving the prominent reduction in antibiotic prescriptions and resistance rates [11, 13]. The government urged in 2017 that all hospitals include stewardship programs in their organizational mandates. In 2017, European Commission issued reports that highlighted the successful implementation of AMS in clinical communities of different countries [14].

4. Need for antimicrobial stewardship programs

Antimicrobial resistance has been increased by the indiscriminate and injudicious use of antibiotics. Bacterial strains are adapted to new resistance mechanisms due to the abuse of sophisticated antibiotics. Bacteria have developed innovative strategies to combat the currently available antibiotics. Many antibiotics have become ineffective due to the modification of penicillin-binding proteins, and the development of beta-lactamases [15]. The term antimicrobial stewardship refers to appropriate measures and a set of coordinated guidelines for the optimization and selection of antimicrobial usage. The professionals should know the duration and type of antibiotics based on pathogenic identification. The chemists, microbiological labs, and infection control agents should work together to guide and assist decision-making persons about antibiotics, dose, and administration routes for maximum output. The global interdisciplinary committee has been trying to optimize the consumption of antibiotics through different antimicrobial stewardship programs to restrict resistance patterns [16].

5. Antimicrobial stewardship strategies

Antimicrobial stewardship consists of two main strategies and the most effective programs typically combine both of them. The first pre-prescription approach includes

Policies	Description
Leadership commitment	Senior management and clinicians who can make decisions regarding the allocation of resources and program implementation.
Accountability and drug expertise	A designated individual or team should be responsible for implementing and managing the antibiotic stewardship program. The team should have expertise in infectious diseases, clinical microbiology, and pharmacology.
Implementing evidence-based policies	Policies and guidelines that reflect evidence-based practices for appropriate antibiotic use. These policies should include recommendations for antibiotic selection, dosing, and duration of therapy.
Antibiotic monitoring and tracking	Antibiotic use and resistance patterns should be monitored to identify areas for improvement and outcomes of interventions should be tracked to evaluate the effectiveness of the stewardship program.
Education and communication	Regular education and communication with healthcare providers, patients, and families are essential to promote appropriate antibiotic use and reduce unnecessary prescribing.
Antimicrobial stewardship interventions	The stewardship program should include interventions such as prospective audits and feedback, clinical guidelines, pathways for antimicrobial order sets, and educational programs for healthcare providers.
Reporting and feedback	Antibiotic use and resistance data should be regularly reported to clinicians and healthcare leaders to promote transparency and accountability for antibiotic prescribing practices.

Table 1.
Core approaches for antimicrobial stewardship programs [17].

the strict authorization and prescription of certain antimicrobials by a selected group of clinicians. These antimicrobial drugs need special approval and a specific time duration which depends on local microbial resistance patterns in different microbial infections. The other post-prescription approach mainly deals with perspective feedback and reviews for prescribed antibiotics to observe the outcomes for dose optimization (**Table 1**). Based on the available microbiology data and the clinical aspects of different cases, the antimicrobial steward analyses current antimicrobial treatments and offers clinicians the to continue, modify, change, or stop the medication [18].

6. Evaluation of antimicrobial stewardship

Patient risk factors and subjectivity should be considered as the major factor for the prescription of antibiotics. Antimicrobial stewardship needs to be followed up and assessed using relevant and acceptable outcome indicators such as intensity of antibiotics consumption, antibiotics resistance and consumption rate, financial condition, and morbidity and mortality rates [19].

Recent meta-analyses and systematic reviews on antimicrobial stewardship policies show the possibility of lowering comparable outcomes in a variety of healthcare settings (**Table 2**).

7. Antimicrobial stewardship programs

The purpose of antimicrobial stewardship programs is to optimize the use of antimicrobial drugs in healthcare facilities such as hospitals, long-term care homes,

Country	Settings	Design	Duration	interventions	Measured results	Reference
China	Secondary and tertiary hospitals	Retrospective observational study	2010–2016	AMS campaign	Antibiotic consumption, prescriptions and expense rate, Resistance rates	[20]
Iran	Tertiary teaching hospitals	Interventional quasi-experimental studies	2017–2019	AMS Educational Programs	Antibiotic prescriptions, the occurrence of CDI, and positive MDR cases	[21]
Germany	The tertiary care teaching hospital	Prospective quasi-experimental, Interrupted time-series study	2014 and 2017	Non-restrictive ASP	Antibiotic prescriptions and cost, mean LOS, CDI incidence rate, and mortality rate in the group of patients who were transferred from emergency rooms to medical wards	[22]
Singapore	Tertiary teaching hospitals	Retrospective cohort studies	2016–2018	Prospective audit and feedback strategy	Following ASP suggestions	[23]
Nigeria	Tertiary hospitals	Pre- and post-intervention study	2016	Development of a protocol, educational meeting, and audit and feedback	Antibiotic consumption, prescriptions, and expense rate and resistance rates	[24]
China	Tertiary hospital	Retrospective observational study	2011 to 2016	Multi-aspect interventions (activity program, educational strategies and performance management)	Antibiotic prescriptions and consumption rate, antibiotic prophylaxis, and resistance rates	[25]

Table 2.
Recent interventions on antimicrobial stewardship programs in hospitals.

and outpatient clinics. These initiatives aim to raise patients' outcomes, control the spread of antibiotic resistance, and cut back on medical expenses.

7.1 Healthcare facility

A list of core components for healthcare facilities has been developed to assist facility management in creating the frameworks required to implement sustainable AMS programs. The core elements are comprised of financial funding, a multidisciplinary dedicated team, health professionals, up-to-date standard treatment recommendations, regular feedback and reviews, and an AMS action plan that sets priorities, evaluates performance, and establishes accountability, collaboration with other programs, a list for healthcare facilities that include a list of restricted antibiotics, basic training for optimal antibiotics use, monitoring quality, susceptibility, types, and resistance rate of antibiotics and medical records and prescription records for the institution (**Figure 2**).

7.2 AMS planning programs

AMS programs should be interlinked with other programs for the optimization of antibiotic usage. A large tertiary hospital with many different specialties will have a larger and more sophisticated AMS program than a local hospital. The deployment of an AMS program is a dynamic and step-by-step process where each facility is built on the infrastructure that already exists [26]. AMS programs conduct situational analysis and faculty action plans based on core elements to identify the missing elements. The situational analysis consists of opportunities, barriers, and enablers at different levels for the participation of clinical professionals. The AMS committee can function independently or it can be integrated into an existing framework such as the infection control, patient safety, drugs and therapeutics committee to define the scope of the work. An infectious disease specialist with interest and experience in infectious illnesses can give expertise to the infection management team. This individual serves as the primary hand for the AMS team for helping with drug prescription, diagnosis, and patient management for the optimization of antibiotics to treat infections [27].

Figure 2.
Core elements of the healthcare facility for antimicrobial stewardship programs.

Patients care should be taken by nurses and expert staff of hospitals because they know the entire first-hand disease history of the patients [28]. This AMS team member should be focused on the timely administration of antibiotics without delaying any dose, therapeutic drug monitoring, high-quality microbiological sampling, and results discussion to promote the accurate prescription of antibiotics. Moreover, nurses should cooperate to monitor the patient's clinical progress, antibiotic consumption intensity, and identify the conditions to shift the route of antibiotics from intravenous (IV) to oral, educate patients, and handle the data collection for feedback and reviews [17].

7.3 Selection and dosing of appropriate antibiotics

The infectious disease workers should recommend the correct duration of treatment with an optimized drug dosage to secure the target more accurately in the shortest period with minimal toxicity. The IV drugs should be shifted to the oral route of administration to shorten the patient's stay at the hospital to avoid catheter-associated infections. The infectious diseases team should follow evidence-based recommendations for the duration of the treatment. The treatment plan for gram-positive and gram-negative bacteria is also established based on the identified pathogen(s) to avoid unnecessary prolonged treatments [29].

Understanding and utilizing pharmacokinetics and pharmacodynamics by hospital and clinical pharmacists is necessary for choosing appropriate antibiotic doses that can help to reduce the use of unnecessary drugs for double therapy. Nursing staff can help explore alternative medicines by accurately documenting and updating antibiotic allergies on a regular basis [30].

8. Antimicrobial stewardship challenges

The antimicrobial stewardship program is challenging due to limited human resources, formal programs, and policies. Antimicrobial stewardship programs require a significant investment of time, money, and resources to implement and maintain.

8.1 Healthcare challenges

The healthcare sector faces major challenges due to lack of equipment, improper infrastructure, large patient burden, and shortage of hospital facilities. These challenges were frequently reported in rural areas due to irrelevant antibiotic prescriptions and malpractices in the healthcare system [31]. In developed countries, hospitals have special antibiotics committees that monitor the antibiotics usage, treatment plans, and evaluate the implementation of hospital's stewardship interventions [32].

8.2 Diagnostic challenges

High burden of infectious diseases has limited the availability of microbiology diagnostic laboratories [33]. There is a dire need for necessary equipment, standard operating protocols, well-trained staff, and quality control systems in diagnostic laboratories [34]. Correct pathogen identification and susceptibility testing are difficult due to the large number of antibiotics that could potentially be tested, the multiple techniques and media needed, and the analysis and validation of unlikely resistance profiles. High testing threshold, lack of experienced microbiologists, and habitat culture media are the main crises due to financial constraints [35].

8.3 Educational challenges

The lack of professional clinical staff has been recorded as the main problem in the path of antimicrobial stewardship practice. In rural areas, antibiotics are prescribed and/or provided in many cases by a diverse group of people including healthcare workers with different learning backgrounds (e.g., nurses, dispensers, chemists, and midwives) as well as local vendors. Many healthcare providers may not have adequate education and training on appropriate antimicrobial use. This can make it difficult for them to make informed decisions about prescribing antimicrobials [36].

Antimicrobial stewardship should be a core element for healthcare workers to fully access over risks of antibiotic misuse and its outcomes for the patients. The development of protocols and guidelines is a mandatory step for the education of medical students and clinical workers. Providers may be educated on the principles of antimicrobial stewardship, appropriate prescribing practices, and the risks associated with inappropriate antibiotic use [13].

8.4 Limited access to antibiotics

Many underdeveloped countries have poor access to different antibiotics in rural and remote areas due to a lack of financial aid, transport, storage, and reliable drug supply system. Limited access to antibiotics can lead to increased morbidity and mortality rates, as well as the spread of infectious diseases [37]. The widespread of non-prescribed antibiotics is another potentially challenging threat for the public sector. Many drugs may be difficult to trace in terms of their sources, quality, and supply chain and the pharmaceutical corporations' influence is frequently neither visible nor regulated [38]. Moreover, US Food and Drug Administration and the European Medicines Agencies have little or no medical access in low- and middle-income countries [39].

9. International guidelines and approaches

The World Health Organization (WHO) has set different roadmaps to monitor the establishment of antimicrobial stewardship policies. The necessity for monitoring and investigation, infection prevention, and control strategies such as immunization and the development of an economic argument for long-term investment in antibiotics have all been mentioned by them as being important components of a complete approach. The availability of affordable and high-quality diagnostic kits and tools, continuous surveillance, and equal distribution of antibiotics is urgently needed as core elements of AMS [12].

The WHO provides a framework for creating National Action Plans (NAP) and analysis systems for a country's progress evaluation. The US Center for Disease Dynamics, Economics and Policy (CDDEP) is funding the establishment of multisectoral national-level working groups in low- and middle-income countries to comprehend and document antibiotic use in human and animal populations in the national context as well as of developing evidence-based interventions [13].

The global action plan extends current efforts on smart drug use and national medicines policy including the use of essential pharmaceutical lists, concerning the regulation of antibiotic availability, quality, and use at the national level. For instance, despite increasing antibiotic resistance in many countries, national essential drug lists in many underdeveloped countries still do not include specific antibiotics such as carbapenems, glycopeptides, or polymyxins. Moreover, programs such as the Global Antibiotic Research and Development Partnership (GARDP), which aims to use partnership models rather than a pharmaceutical industry-driven strategy to create novel antibiotic treatments may be capable of ensuring ethical use and unlimited access to different antibiotics [40].

Targeted strategies such as shifting from intravenous to dosage form, improving surgical antibiotic prophylaxis, and introducing standard antibiotic formulations might be taken into consideration. Improving antibiotic hang time is another easy approach to improving patients' outcomes [41].

10. Case studies

Hospital-based antimicrobial stewardship programs (ASPs) use several coordinated strategies to optimize the use of antibiotics. Verma et al., [42] have reported pre- and post-intervention study designs that compared the outcomes of AMS on patients admitted from November 2017 to January 2018. Daily assessments and feedbacks, restrictions on the use of antibiotics, daily bedside evaluations, and educational activities to raise awareness among residents and nurses were part of the ASP interventions. The study involved the prescription of 1331 antibiotics to 695 patients. Prophylactic antibiotic use was decreased by 11% (p = 0.001) in this period. In terms of length, choice, indication, and method of administration, the prescription pattern was considerably improved in the intervention phase compared to the pre-intervention phase by 8%, 14%, 2%, and 8%, respectively. The outcomes of this study indicated positive feedback on the implementation of AMS in hospital units for the optimization of antibiotic dosage.

Jones et al., [43] presented the hypothesis of the positive relationship of professional healthcare workers with the success of AMS. During six months intervention period, antibiotic consumption was lowered by 6.5% compared to previous years' reports. A daily review of antibiotics prescription and administration documentation was significantly improved.

Similarly, [44] conducted telehealth-based antimicrobial stewardship programs in two hospitals. In six months duration, 1419 proposals were presented and 1262 (88.9%) of them were followed through with a 24.4% decrease in broad-spectrum antibiotic consumption. Antimicrobial expenses were estimated to save $1,42,629.83 annually. An intensive ASP strategy enabled in the community hospital context *via* telehealth resulted in a lower broad-spectrum antibiotic usage and minimum antimicrobial expenses.

11. Conclusion

The rising frequency of antibiotic resistance in hospitals as well as communities is an urgent issue as hospitalized patients become more difficult to treat. To address the problem(s) of AMR, it is important to promote the appropriate use of antimicrobial drugs, develop new and more effective drugs, and improve infection prevention and control practices in healthcare settings and the community. Antimicrobial stewardship practices, concepts, and interventions are the key stages in the prevention and control of antimicrobial resistance. Antimicrobial stewardship may offer every professional the tools they need to use valuable resources.

Author details

Hafiza Salaha Mahrosh[1] and Ghulam Mustafa[2]*

1 Department of Biochemistry, University of Agriculture Faisalabad, Faisalabad, Pakistan

2 Department of Biochemistry, Government College University Faisalabad, Faisalabad, Pakistan

*Address all correspondence to: drghulammustafa@gcuf.edu.pk

References

[1] Dyar O et al. What is antimicrobial stewardship? Clinical Microbiology and Infection. 2017;**23**(11):793-798

[2] Schuts EC et al. Current evidence on hospital antimicrobial stewardship objectives: A systematic review and meta-analysis. The Lancet Infectious Diseases. 2016;**16**(7):847-856

[3] Ababneh MA, Nasser SA, Rababa'h AM. A systematic review of antimicrobial stewardship program implementation in middle eastern countries. International Journal of Infectious Diseases. 2021;**105**: 746-752

[4] Barlam TF et al. Implementing an antibiotic stewardship program: Guidelines by the Infectious Diseases Society of America and the Society for Healthcare Epidemiology of America. Clinical Infectious Diseases. 2016; **62**(10):e51-e77

[5] Nadeem SF et al. Antimicrobial resistance: More than 70 years of war between humans and bacteria. Critical Reviews in Microbiology. 2020;**46**(5): 578-599

[6] Nathwani D et al. Value of hospital antimicrobial stewardship programs [ASPs]: A systematic review. Antimicrobial Resistance & Infection Control. 2019;**8**:1-13

[7] Dadgostar P. Antimicrobial resistance: Implications and costs. Infection and drug resistance. 2019;**12**: 3903-3910

[8] MacDougall C, Polk RE. Antimicrobial stewardship programs in health care systems. Clinical Microbiology Reviews. 2005;**18**(4): 638-656

[9] Pollack LA, Srinivasan A. Core elements of hospital antibiotic stewardship programs from the Centers for Disease Control and Prevention. Clinical Infectious Diseases. 2014;**59** (suppl_3):S97-S100

[10] Rosenblatt-Farrell N. The landscape of antibiotic resistance. Environmental Health Perspectives. 2009;**117**(6):A244-A250

[11] Davey P, et al. Interventions to improve antibiotic prescribing practices for hospital inpatients. Cochrane Database of Systematic Reviews. 2017;**2**: 14651858

[12] World Health Organization. Global action plan on antimicrobial resistance. Geneva, Switzerland: World Health Organization; 2015. ISBN 978 924 1509763

[13] Cox JA et al. Antibiotic stewardship in low-and middle-income countries: The same but different? Clinical Microbiology and Infection. 2017;**23**(11): 812-818

[14] Majumder MAA et al. Antimicrobial stewardship: Fighting antimicrobial resistance and protecting global public health. Infection and drug resistance. 2020;**13**:4713-4738

[15] Rice LB. Antimicrobial stewardship and antimicrobial resistance. Medical Clinics. 2018;**102**(5):805-818

[16] Sahra S, Jahangir, De Chavez V. Antimicrobial stewardship: A review for internal medicine physicians. Cureus. 2021;**13**(4):e14385

[17] World Health Organization. Antimicrobial Stewardship Programmes

in Health-Care Facilities in Low- and Middle-Income Countries. Geneva, Switzerland: A Who Practical Toolkit; 2019. ISBN 978-92-4-151548-1

[18] Doron S, Davidson LE. Antimicrobial stewardship. Mayo Clinic Proceedings. 2011;**86**(11):1113-1123

[19] Horikoshi Y et al. Sustained pediatric antimicrobial stewardship program with consultation to infectious diseases reduced carbapenem resistance and infection-related mortality. International Journal of Infectious Diseases. 2017;**64**: 69-73

[20] Xiao Y et al. Change in antibiotic use in secondary and tertiary hospitals nationwide after a national antimicrobial stewardship campaign was launched in China, 2011–2016: An observational study. The Journal of Infectious Diseases. 2020;**221**(Supplement_2):S148-S155

[21] Mardani M, Abolghasemi S, Shabani S. Impact of an antimicrobial stewardship program in the antimicrobial-resistant and prevalence of clostridioides difficile infection and amount of antimicrobial consumed in cancer patients. BMC Research Notes. 2020;**13**(1):1-5

[22] Savoldi A et al. Impact of implementing a non-restrictive antibiotic stewardship program in an emergency department: A four-year quasi-experimental prospective study. Scientific Reports. 2020;**10**(1):1-8

[23] Wee LE et al. Who listens and who doesn't? Factors associated with adherence to antibiotic stewardship intervention in a Singaporean tertiary hospital. Journal of Global Antimicrobial Resistance. 2020;**22**:391-397

[24] Abubakar U, Syed Sulaiman SA, Adesiyun AG. Impact of pharmacist-led antibiotic stewardship interventions on compliance with surgical antibiotic prophylaxis in obstetric and gynecologic surgeries in Nigeria. PLoS One. 2019; **14**(3):e0213395

[25] Wang H et al. Impact of antimicrobial stewardship managed by clinical pharmacists on antibiotic use and drug resistance in a Chinese hospital, 2010–2016: A retrospective observational study. BMJ Open. 2019; **9**(8):e026072

[26] Mendelson M et al. How to start an antimicrobial stewardship programme in a hospital. Clinical Microbiology and Infection. 2020;**26**(4):447-453

[27] Goff DA, Rybak MJ. Global antimicrobial stewardship: challenges and successes from frontline stewards. 2015;**4**(Suppl 1):S1S3

[28] Brink AJ et al. Passing the baton to pharmacists and nurses: New models of antibiotic stewardship for South Africa?: Guest editorial. South African Medical Journal. 2016;**106**(10):947-948

[29] Sampathkumar P. Reducing catheter-associated urinary tract infections in the ICU. Current Opinion in Critical Care. 2017;**23**(5):372-377

[30] Satterfield J, Miesner A, Percival K. The role of education in antimicrobial stewardship. Journal of Hospital Infection. 2020;**105**(2):130-141

[31] Zhang C et al. The professional status of infectious disease physicians in China: A nationwide cross-sectional survey. Clinical Microbiology and Infection. 2018;**24**(1):82. e5-82. e10

[32] Howard P et al. An international cross-sectional survey of antimicrobial stewardship programmes in hospitals.

Journal of Antimicrobial Chemotherapy. 2015;**70**(4):1245-1255

[33] Feikin DR et al. The burden of common infectious disease syndromes at the clinic and household level from population-based surveillance in rural and urban Kenya. PLoS One. 2011;**6**(1): e16085

[34] Barbé B et al. Implementation of quality management for clinical bacteriology in low-resource settings. Clinical Microbiology and Infection. 2017;**23**(7):426-433

[35] Om C et al. “If it'sa broad spectrum, it can shoot better”: Inappropriate antibiotic prescribing in Cambodia. Antimicrobial Resistance & Infection Control. 2016;**5**(1):1-8

[36] Morgan DJ et al. Non-prescription antimicrobial use worldwide: A systematic review. The Lancet Infectious Diseases. 2011;**11**(9):692-701

[37] Pulcini C et al. Ensuring universal access to old antibiotics: A critical but neglected priority. Clinical Microbiology and Infection. 2017;**23**(9):590-592

[38] Parry J. Former Head of GSK China Is Charged with Bribery. British Medical Journal Publishing Group; 2014

[39] Collignon P, et al. Antimicrobial resistance: the major contribution of poor governance and corruption to this growing problem. PloS one. 2015;**10**(3): e0116746

[40] World Health Organization. WHO Model List of Essential Medicines, 20th List (March 2017, Amended August 2017). Geneva, Switzerland: World Health Organization; 2017

[41] Messina AP, van den Bergh D, Goff DA. Antimicrobial stewardship with pharmacist intervention improves timeliness of antimicrobials across thirty-three hospitals in South Africa. Infectious Diseases and Therapy. 2015;**4**: 5-14

[42] Verma M et al. Antimicrobial stewardship programme in a trauma Centre of a tertiary care hospital in North India: Effects and implementation challenges. Journal of Global Antimicrobial Resistance. 2019;**17**: 283-290

[43] Jones AS, et al. Impact of positive feedback on antimicrobial stewardship in a pediatric intensive care unit: A quality improvement project. Pediatric quality & safety. 2019;**4**(5):e206

[44] Shively NR et al. Impact of a telehealth-based antimicrobial stewardship program in a community hospital health system. Clinical Infectious Diseases. 2020;**71**(3):539-545

Chapter 2

Antibiotic Stewardship: How It Is Implemented in Primary Healthcare Facility

Rini S. Handayani and Vita Pertiwi

Abstract

Antibiotic stewardship plays an important role in controlling antibiotic resistance. The problem of antibiotic resistance in primary healthcare has not been given much attention, so far, the focus has been on the hospital. Antibiotic resistance control needs to start from primary healthcare facilities such as community health centers, clinics, and independent doctor practices because patients who enter the hospital are often already resistant. Therefore, it is necessary to identify things that can be done to control antibiotic resistance in primary health care facilities. Things that can be done include making policies or regulations that support antibiotic stewardship in primary healthcare, optimizing available facilities and infrastructure, improving facilities that support antibiotic stewardship, increasing capable human resources, monitoring and evaluating antibiotic prescribing, and building system information on antibiotic resistance that is easily accessible to health workers in primary health care facilities.

Keywords: antibiotic stewardship, primary healthcare, resistance policy, primary care provider, stewardship implementation

1. Introduction

According to World Health Organization (WHO), antibiotic stewardship is a comprehensive integrated series of activities to promote the responsible and appropriate use of antibiotics so that patients get optimal treatment results. These strategies are reducing antibiotic utilization by increasing the rational use of antibiotics and reducing the incidence of infectious diseases by carrying out immunizations. The irrational use of antibiotics includes inaccuracies in diagnosis, dosage, interval, and duration. Prescribing antibiotics recklessly is a predisposing factor for the emergence of antibiotic resistance. In primary health facilities, antibiotics are widely prescribed for acute Upper Respiratory Tract Infections (URTI), even though URTI is mostly caused by viruses and antibiotic is not a proper drug choice. Using broad-spectrum antibiotics too frequently is also not recommended. The existence of guidelines for prescribing appropriate antibiotics with the application of evidence-based interventions helps clinicians to use antibiotics wisely. Besides, the proper management of antibiotics also plays a role in controlling antibiotic-resistant bacteria and saving costs. Therefore, doctors and pharmacists in primary health facilities play

IntechOpen

an important role in overcoming inappropriate antibiotic prescribing problems and antibiotic stewardship is a key strategy to maintain the effectiveness of antibiotics in the future [1–9]. Implementation of antibiotic stewardship will be effective if it is carried out cross-sectorally from the national level to the regional level by involving all parties such as local health offices, hospitals, primary health facilities, health workers like doctors, pharmacists, and the community.

2. Problems

Currently, the antibiotic stewardship strategy is mostly carried out in hospitals, especially referral hospitals, rather than primary health facilities [1]. In contrast to referral hospitals, the primary health center has limited human resources, facilities, and infrastructure. It is estimated that about 80% of antimicrobials are consumed in primary health facilities. Therefore, antibiotic stewardship must be started in primary health facilities immediately. Limited sources should not be used as a barrier to initiating antibiotic stewardship. Although it is not yet ideal, it is hoped that antibiotic stewardship in primary health facilities will play an important role in controlling antibiotic resistance. Besides limited resources, the understanding of antibiotic resistance among health workers is not comprehensive. Lack of risk awareness of antibiotic resistance among health workers like doctors and pharmacists is also one of the leading factors for increasing antibiotic resistance. Therefore, interventions to increase the awareness of doctors and pharmacists in primary health centers about using antibiotics wisely need to be carried out. This intervention will help increase rational antibiotic utilization. Other factors include physician non-compliance with antibiotic utilization management, limited access to microbiological testing to support diagnosis, lack of support and coordination with local policymakers, and limited access to antibiotic resistance data in the community [5, 10]. Even though policymakers have made several interventions by developing antibiotic indication and utilization regulation, for example in the case of URTI, the rate of inappropriate antibiotic prescribing is still high [3]. Optimal antibiotic stewardship implementation is expected to reduce inappropriate antibiotic prescribing. Antibiotic stewardship programs and activities can be successful if the challenges faced by primary health facilities can be overcome [3, 7]. Therefore, challenges or obstacles to the implementation of antibiotic stewardship in primary health facilities must be minimized, considering that inappropriate antibiotic prescribing often occurs in primary health facilities.

3. Antibiotic stewardship implementation in primary healthcare facilities

3.1 Antibiotic stewardship policies in primary health care facility

In developing countries, enforcement of policies prohibiting the free sale of antibiotics is often not optimal. This allows people to self-medicate using antibiotics that are not needed [11]. Improvements across the health system and drug supply chain are needed to tackle the problem of antibiotic resistance [9]. In addition, the government needs to make policies to strengthen the role of primary healthcare facilities in controlling antibiotic resistance because they are the ones dealing with the most patients. Antibiotic stewardship policy needs to start from the upstream to primary healthcare facilities to have a significant impact. Antimicrobial stewardship programs in the form of education, clinical support, supervision, and policy have reduced rates of prescription writing and inappropriate prescribing [12]. The government needs to immediately make national policies related to antibiotics following WHO recommendations, including the Access, Watch, and Reserve (AWaRe) classification in national guidelines for antibiotics application [13].

The Ministry of Public Health in Qatar developed the NAP (National Plan of Action) to Fight Antimicrobial Resistance (AMR) in collaboration with the WHO Regional Office for the Eastern Mediterranean (WHO/EMRO)) [14]. Indonesia, through the Ministry of Health, has established rational drug prescription indicators for URTI and diarrhea at Public Health Centers. Tolerance to the use of antibiotics in URTI is at 20% and in diarrhea is at 8%. This is based on the fact that antibiotics are often given to treat diarrhea and acute respiratory infections, although in most cases these conditions are caused by viruses and do not require antibiotics [9]. This target is stated in the Regulation of the Coordinating Minister for Human Development and Culture of the Republic of Indonesia number 7 of 2021 concerning the national action plan for controlling antimicrobial resistance for 2021–2024. The Indonesian Ministry of Health has also made pharmacists as the agents of change which are part of the Community Movement to Use Drug Intelligently program. One of the goals of this program is to educate health workers and the public to use antibiotics wisely and be aware of the antibiotic resistance dangers. Another important component of the program is comprehensive surveillance of antibiotic prescription and antimicrobial resistance. This data is intended to help antibiotic stewardship efforts be more precise and wise [12].

3.2 Optimizing primary healthcare's facilities to support the antibiotic stewardship program

In developing countries, facilities and infrastructure in primary healthcare facilities are often limited, especially in rural and remote areas. There is a shortage of diagnostic tools so that the diagnosis becomes inaccurate and inappropriate antibiotic prescribing occurs when the patient does not need antibiotics [11]. Rapid diagnostic tests will allow doctors to prescribe specific antibiotics rather than broad-spectrum antibiotics and to differentiate between viral and bacterial infections [9].

The lack of human resources, the inaccessibility of diagnostic tests, and the cheap price of antibiotics resulted in a tendency to give treatment with empiric antibiotics directly without conducting cultures because they were considered more efficient economically [9, 15]. General practitioners in developed countries commonly use the rapid antigen detection test for the diagnosis of streptococcal pharyngitis. The C-reactive protein (CRP) kit is used to differentiate serious respiratory infections [16].

In addition to the absence of a rapid diagnostic test, inappropriate application of guidelines, unavailability of specific antibiogram data, and the short consultation time due to many patients are also factors triggering the irrational use of antibiotics [14]. Public understanding of the dangers of antibiotic resistance plays an important role in controlling the spread of antibiotic resistance. Unfortunately, people often believe that antibiotics will help cure colds and flu caused by viruses. Whereas viral infections generally do not require antibiotics [11].

Poor sanitation in primary healthcare facilities also facilitates spreading of the microbes that are resistant to antibiotics. To control the spread of antibiotic-resistant microbes in the environment, primary healthcare facilities must be clean. Contamination in water supplies is driving the spread of antibiotic-resistant microbes. This clean primary healthcare can be achieved if it is supported by community health workers who are competent in environmental sanitation. Adequate water, good sanitation, and public health workers implementing effective infection prevention and control procedures will control the spread of antibiotic-resistant microbes. Unfortunately, many primary health care facilities do not follow basic sanitation measures such as aseptic technique, hand washing, and the use of personal protective equipment, such as gloves and masks. Without clean water and good sanitation in primary healthcare facilities, they become infection exposure points for patients and the staff. These inadequate facilities can be the starting point for transmission of antimicrobial resistance in the community [9].

A comprehensive assessment of primary healthcare facilities in resource-limited settings is essential to strengthening the Infection Prevention and Control (IPC) measures and Antibiotic Stewardship Programs (ASP) activities. This approach for assessing existing IPC and ASP activities will provide relevant data for establishing further policies regarding adequate facilities and infrastructure in primary healthcare facilities to support the Antibiotic Stewardship Program [17].

3.3 Increase health worker's capacity in primary healthcare

Some health workers consider antibiotic resistance as a national and even international problem but sadly this does not affect their daily practice in primary healthcare facilities. The others do not understand and have full awareness about antibiotic resistance. Some doctors in primary healthcare facilities may prescribe antibiotics out of habit or are reluctant to follow the new policy in prescribing antibiotics. Therefore, antibiotic prescribing varies greatly [9, 18]. Other health workers like pharmacists and informal health service providers, especially in countries and regions where there is relatively free access to antibiotics also contribute to antibiotic resistance [12]. Raising awareness and understanding the antibiotic resistance must start with professional education, and continue to any other advanced education program. This continuing education is carried out with the hope of behavior changes in health workers [16]. Besides educational interventions, it is necessary to audit and give feedback because it will increase the rational use of antibiotics. If needed, policy changes and an information system are formed to remind and promote the rational use of antibiotics in primary healthcare facilities [10, 18].

In Indonesia, especially in rural and remote areas, the prescription of antibiotics is also carried out by nurses or midwives due to the absence of general practitioners. Therefore, education interventions about the dangers of antibiotics should also be given to nurses or midwives. Unfortunately, continuing education for health workers in primary healthcare facilities is not yet intensive. Poor management of antibiotics in primary healthcare facilities also has a role in antibiotic resistance [7].

In addition to the health workers mentioned above, pharmacists also play a role in controlling antibiotic resistance. Pharmacists have a significant role in ensuring the proper consumption of antibiotics. Unfortunately, pharmacists often lack the education about advanced antibiotic resistance even though have an important role in the Antibiotic Stewardship Program. Another crucial point is the lack of collaboration between health workers such as doctors, nurses, midwives, and pharmacists [19].

Apart from health workers, understanding and being aware of the antibiotic resistance danger is also crucial for patients. The public also needs to be educated regarding this matter. Unawareness of the public regarding the proper use of antibiotics and free access to antibiotics is a potential risk for antibiotic resistance [10–12]. People

DOI: http://dx.doi.org/10.5772/intechopen.113102

believe that antibiotics can cure diseases caused by viral infections such as the common cold or flu [11]. People often self-medicate using leftover antibiotics that have been previously prescribed for diseases that are not necessarily caused by bacteria. Non-compliance with the dosage and rules of use that doctors have prescribed is one of the factors for the development of antibiotic resistance in a community [12]. This behavior needs to be changed, they must be taught to only use antibiotics according to the doctor's instructions and not self-medicate using antibiotics [18]. Effective education and communication to the public to increase public awareness is part of the Antibiotic Stewardship Program [7]. Public education about antibiotic utilization and the dangers of antibiotic resistance can be carried out through talk shows with pharmacists and doctors through electronic media such as TV, radio, YouTube, and other mass media [16]. Besides educating health workers, raising awareness and educating the public about antibiotic resistance also needs to be done.

Although doctors are the most influential health professionals in addressing the problem of antibiotic resistance in primary healthcare facilities, treatment decisions are sometimes influenced by patient demands and are often driven by an unfounded belief in getting the 'quickest cure' with antibiotics [10–12]. Inappropriate use of antibiotics and demands on doctors to prescribe antibiotics can result from a poor understanding of the implications of antibiotic abuse, limited access to trained doctors, limited resources, and health literacy [12]. General practitioners often feel that patients with URTI consult with the hope of getting antibiotics [18]. Even though doctors understand the dangers of antibiotic resistance, sometimes doctors prescribe antibiotics to maintain good relations with patients and avoid patient dissatisfaction. This paradox still occurs and may be one of the main contributors to the continued over-prescribing of antibiotics for URTI [11, 18]. The gap in knowledge about antibiotic resistance that contributes to the continuation of this antibiotic resistance needs to be eliminated. Therefore, better communication skills are needed for patients and the wider community. Continuing education materials apart from material on the rational use of antibiotics, the dangers of antibiotic resistance are also material on communicative ways of communicating to patients and the wider community. Communicative communication with patients is the key to making joint decisions to use antibiotics rationally. This has been shown to reduce antibiotic prescriptions for URTI in several studies [18, 20].

Given a lot of factors cause antibiotic resistance, the high risk of antibiotic resistance, and the involvement of many parties, this antibiotic resistance must be overcome by all parties in collaboration. Working together to tackle antibiotic resistance, both from the sector in humans and livestock, (one health priority agenda) will be more successful than doing it alone [11, 21]. Doctors and pharmacists in primary healthcare can be the spearheads of the "one health priority agenda" because primary healthcare does not only provide health services but also has the functions and duties of carrying out health promotion and preventive programs.

3.4 Antibiotic monitoring and evaluation

Monitoring and evaluation of antibiotic prescribing and resistance are very important in controlling antibiotic resistance. This is necessary as a feedback basis for primary healthcare facilities, an intervention that needs to be carried out by the local health authority (public health office) or on a national scale by the Ministry of Health. The results of monitoring and evaluation are also needed as a reference for improving antibiotic prescribing guidelines.

Comprehensive surveillance of antibiotic prescription and resistance is an important component of the Antibiotic Stewardship Program [12]. WHO issued the "Global Antimicrobial Resistance and Use Surveillance System (GLASS)" program on October 22, 2015. This GLASS program is the first global collaborative effort to standardize AMR surveillance. Surveillance is a way to obtain data as a basis for assessing the spread of AMR and to inform and monitor the impact of local, national, and global strategies. Based on the surveillance results, infection prevention, control responses, and policy improvements are determined. GLASS was compiled by combining data from AMR monitoring in humans, namely monitoring of microbial resistance and antimicrobials utilization including in the food chain (use of antimicrobials in animal husbandry and fisheries) and resistant microbes in the environment [22]. Unfortunately, this can only be done at the hospital level, it cannot be done yet in primary health facilities due to the absence of microbiological examination facilities and limited infrastructure, as well as health human resources to carry out antimicrobial resistance tests. Collaboration with hospitals is needed to overcome this problem. Besides the surveillance program, data on antibiotic consumption at the primary healthcare is also needed for monitoring and evaluating antibiotic management. Community-level resistance and consumption data are needed to monitor resistance, inform treatment guidelines and patient care, and understand prescribing patterns. For evaluating and monitoring antibiotic utilization, WHO classifies antibiotics into three groups, namely Access, Watch, and Reserve. The Access group is a group of antibiotics that have activity against a wide range of commonly encountered susceptible bacteria while also having lower resistance potential than other groups. This group of antibiotics is recommended as the first or second choice of infectious disease treatment. Therefore, the Access group should be the most widely available antibiotics in primary healthcare [9]. There is a direct correlation between antibiotic use and resistance. High consumption of antibiotics leads to high levels of resistance [16]. In Indonesia, monitoring of antibiotic use in primary healthcare facilities is carried out on the use of antibiotics in non-pneumonic URTI and non-specific diarrhea. The Indonesian government has also established a national formulary that includes a list of antibiotics used in primary healthcare facilities. Community Health Centers were asked to report the use of antibiotics in non-pneumonic URTI and non-specific diarrhea to the local public health office and then they will report to the Ministry of Health, meanwhile, an evaluation of the suitability types of antibiotics prescribed at the primary healthcare with the national formulary had not been carried out.

3.5 Building an information system on antibiotic resistance

An obstacle in making the right diagnosis in primary healthcare facilities is the absence of microbiological examination to determine whether the cause of infection is bacteria or virus. This situation makes choosing precise antibiotics harder. Microbiological examination currently only exists in hospitals, so collaboration with hospitals is needed. Building an antibiotic-resistant information system that is easily accessible to health workers in primary healthcare facilities can be a solution to overcome this [9].

An information system coordinated by the local health authority (public health office) in collaboration with hospitals is urgently needed considering that the hospital has antibiotic resistance data as well as laboratory facilities and human resources for the microbiology examination laboratory. Primary healthcare facilities will provide information on the use of antibiotics. A comprehensive information system involving hospitals, primary healthcare facilities, and health offices will be effective in increasing rational antibiotic prescribing [9].

Data on antibiotic utilization and antibiotic resistance that are integrated into a system can monitor antibiotic use and emerging resistance so that future antibiotic prescribing can be based on local resistance patterns and results with more accurate diagnosis and therapy. This information system can also be accessed by primary healthcare facilities in rural areas, given that internet access is now quite extensive in rural areas. This will overcome the problem of health workers in rural areas who lack information about local resistance patterns. This application should not be too complicated so that it can be accessed by smartphone. An app launched in South Africa in late 2017, provides information to patients about their diagnosis, proper use of antimicrobials, and infection prevention; physicians can enter patient characteristics and symptoms to obtain guidance on their diagnosis and local prevalence of resistance [9].

This information system can also be used by the health office to monitor and provide feedback to primary healthcare facilities. This comprehensive information will assist doctors in prescribing appropriate antibiotics or not prescribing antibiotics for patients with viral infections who usually do not require antibiotics [9]. Information systems that are supported by interventions can provide a better result of antibiotic resistance prevention, but this information and interventions are very specific regionally, so they will vary depending on the region [10].

In addition to the information system intended for health workers, the public health office can create an information system intended for the community. This system will be useful for socializing about the danger of antibiotic resistance with interesting content for the public. The Internet can be one of the media because it is widely accessed by the public. There were an estimated 4.90 billion social media users worldwide in 2017 and is projected to increase to 5.85 billion in 2027 [23]. Public health offices can promote proper use of antibiotics, and hygiene practices to the public as part of antibiotic stewardship activity. Besides, this information system can be used against media that provide wrong information about health problems, such as the use of antibiotics to treat colds and influenza and the dangers of vaccines [9].

4. Conclusion

An Antibiotic Stewardship Program is necessary to solve antibiotic resistance. Focus programs need to change from hospital-based to primary healthcare-based due to their large coverage. An implementation that can be done in primary healthcare is forming a solid policy, optimizing its infrastructure, upgrading human resources, monitoring and evaluating antibiotic use routinely, and building an information system. Cooperation between primary healthcare, hospitals, and public health offices made a comprehensive and powerful antibiotic stewardship program.

Conflict of interest

The authors declare no conflict of interest.

Author details

Rini S. Handayani[1] and Vita Pertiwi[2*]

1 National Research and Innovation Agency, Jakarta, Indonesia

2 Prambanan Public Hospital, Yogyakarta, Indonesia

*Address all correspondence to: vitapertiwi96@gmail.com

References

[1] WHO. Policy Guidance on Integrated Antimicrobial Stewardship Activities. Geneva: World Health Organization; 2021. Licence: CC BY-NC-SA 3.0 IGO

[2] WHO. Antimicrobial Stewardship Programmes in Health-Care Facilities in Low- and Middle-Income Countries. A Practical Toolkit. Geneva: World Health Organization; 2019. Licence: CC BY-NC-SA 3.0 IGO

[3] Avent ML, Cosgrove SE, Price-Haywood EG, et al. Antimicrobial stewardship in the primary care setting: From dream to reality? BMC Family Practice. 2020;**21**(134):1-9. DOI: 10.1186/s12875-020-01191-0

[4] Barlam TF, Cosgrove SE, Abbo LM, et al. Implementing an antibiotic stewardship program: Guidelines by the Infectious Diseases Society of America and the Society for Healthcare Epidemiology of America. Clinical Infectious Diseases. 2016;**62**(10):51-77. DOI: 10.1093/cid/ciw118

[5] WHO. Antimicrobial Stewardship Interventions: A Practical Guide. Copenhagen: WHO Regional Office for Europe; 2021. Licence: CC BY-NC-SA 3.0 IGO

[6] Aslam B, Wang W, Arshad MI, et al. Antibiotic resistance: A rundown of a global crisis. Infection and Drug Resistance. 2018;**11**:1645-1658. DOI: 10.2147/IDR.S173867

[7] Engler D, Meyer JC, Schellack N, et al. Antimicrobial stewardship activities in public healthcare facilities in South Africa: A baseline for future direction. Antibiotics (Basel). 2021;**10**(8):996. DOI: 10.3390/antibiotics10080996

[8] Gi Mancuso G, Midiri A, Gerace E, et al. Bacterial antibiotic resistance: The most critical pathogens. Pathogens. 2021;**10**(10):1310. DOI: 10.3390/pathogens10101310

[9] WHO. Antimicrobial Resistance and Primary Health Care, Technical Series on Primary Health Care. Geneva: World Health Organization; 2018. Licence: CC BY-NC-SA 3.0 IGO

[10] Yao L, Yin J, Huo R, et al. The effects of the primary health care providers' prescription behavior interventions to improve the rational use of antibiotics: A systematic review. Global Health Research and Policy. 2020;**5**:45. DOI: 10.1186/s41256-020-00171-2

[11] Dadgostar P. Antimicrobial resistance: Implications and costs. Infection and Drug Resistance. 2019;**12**:3903-3910. DOI: 10.2147/IDR.S234610

[12] Yau JW, Thor SM, Tsai D, et al. Antimicrobial stewardship in rural and remote primary health care: A narrative review. Antimicrobial Resistance & Infection Control. 2021;**10**(105):1-33. DOI: 10.1186/s13756-021-00964-1

[13] Sulis G, Adam P, Nafade V, et al. Antibiotic prescription practices in primary care in low- and middle-income countries: A systematic review and meta-analysis. PLoS Medicine. 2020;**17**(6):1-20. DOI: 10.1371/journal.pmed.1003139

[14] Sharaf N, Al-Jayyousi GF, Radwan E, et al. Barriers of appropriate antibiotic prescription at PHCC in Qatar: Perspective of physicians and pharmacists. Antibiotics (Basel). 2021;**10**(3):317. DOI: 10.3390/antibiotics10030317

[15] Chinemerem Nwobodo D, Ugwu MC, Oliseloke Anie C, et al. Antibiotic resistance: The challenges and some emerging strategies for tackling a global menace. Journal of Clinical Laboratory Analysis. 2022;**36**(9):1-10. DOI: 10.1002/jcla.24655

[16] Llor C, Bjerrum L. Antimicrobial resistance: Risk associated with antibiotic overuse and initiatives to reduce the problem. Therapeutic Advances in Drug Safety. 2014;**5**(6):229-241. DOI: 10.1177/2042098614554919

[17] Harun MGD, Anwar MMU, Sumon SA, et al. Rationale and guidance for strengthening infection prevention and control measures and antimicrobial stewardship programs in Bangladesh: A study protocol. BMC Health Services Research. 2022;**22**(1):1-11. DOI: 10.1186/s12913-022-08603-0

[18] Hansen MP, Hoffmann TC, McCullough AR, et al. Antibiotic resistance: What are the opportunities for primary care in alleviating the crisis? Frontiers in Public Health. 2015;**3**(35):1-5. DOI: 10.3389/fpubh.2015.00035

[19] Luhraimaduretnoasvinigita B, Kristina S, Ika PS. Antibiotics stewardship practice among community pharmacists in Indonesia: A cross-sectional survey. International Journal of Pharmaceutical Research. 2020;**11**(4):176-180. DOI: 10.31838/ijpr/2019.11.04.028

[20] Medina-Perucha L, García-Sangenís A, Moragas A, et al. Autonomy, power dynamics and antibiotic use in primary healthcare: A qualitative study. PLoS One. 2020;**15**(12):1-21. DOI: 10.1371/journal.pone.0244432

[21] WHO. A One Health Priority Research Agenda for Antimicrobial Resistance. Geneva: World Health Organization, Food and Agriculture Organization of the United Nations, United Nations Environment Programme and World Organisation for Animal Health; 2023. Licence: CC BY-NC-SA 3.0 IGO

[22] WHO. Global Antimicrobial Resistance and Use Surveillance System (GLASS) [Internet]. 2022. Available from: https://www.who.int/initiatives/glass [Accessed: August 12, 2023]

[23] Ruby D. Social Media Users in 2023 (Global Demographics) [Internet]. 2023. Available from: https://www.demandsage.com/social-media-users/#:~:text=Social%20Media%20Users%202023%20(Top,base%20with%202.99%20billion%20users [Accessed: August 01, 2023]

Chapter 3

Applications of Antimicrobial Stewardship and Natural Product Chemistry in Tackling Antimicrobial Resistance

Khalifa Musa Muhammad and Mansurat Oluwatoyin Shoge

Abstract

Antimicrobial resistance (AMR) is a major concern for global health security because of its impact on human, environment, and animal health. This tendency of AMR was corroborated by Alexander Fleming who discovered the first antibiotic. This chapter focuses on the global concern of AMR, its causes, and solutions. Antimicrobial stewardship (AMS) is one of the solutions employed globally to tackle the challenge of AMR. The objective of the AMS includes: reducing antibiotic abuse, lowering healthcare costs, and tackling AMR. Therefore, it is pertinent to decrease AMR and protect global health. Many countries are implementing antimicrobial stewardship programs (ASPs) in order to reduce AMR. The misuse of antibiotics is one of the major factors that cause AMR. To reduce antibiotic abuse pharmacists have a key role to play. Finding new drugs to treat resistant pathogens is another solution to AMR. Plants have contributed immensely to traditional medicine and drug discovery due to the presence of bioactive secondary metabolites. They have the potential to contribute immensely to tackling AMR.

Keywords: antimicrobial resistance, antimicrobial stewardship, global health, plants, secondary metabolites

1. Introduction

Antibiotics are considered as a substance of biological origin or produced by a microorganism that can be lethal to other organisms or inhibit their growth [1]. In 1928 Alexander Fleming discovered the first antibiotic from fungi and by 1940 other useful antibiotics were discovered from bacteria [2]. The period of 1930–1960 is regarded as the antibiotics golden era due to the discovery of many antibiotics [3].

The drugs that are most prescribed across the globe are antibiotics [4]. It is estimated that by 2030 the use of antibiotics will increase up to 67% in highly populated countries around the world [5]. Antibiotics have contributed immensely by reducing mortality and morbidity rates, especially in developing countries [6]. However, there is a need for new antibiotics due to the existence of resistant bacteria and the advent of new diseases [2].

Antibiotics mechanism of action includes:

- Inhibiting the synthesis of cell wall
- Breaking down of cell membrane function or structure
- Inhibiting the structure and function of nucleic acids
- Inhibiting protein synthesis
- Blocking of important metabolic pathways [1].

Antibiotics have played an unparalleled role in the advancement of society and medicine, and they are now a requirement in all healthcare systems. Antibiotic treatment has contributed to controlling bacterial infections; enhancing major surgeries; cancer therapy; and other successful aspects of modern medicine [5].

It is pertinent to sustain the effectiveness of present antibiotics in order to prevent any retrogression recorded in dialysis, surgery, and chemotherapy among others [7]. If the challenge of antimicrobial resistance (AMR) is not tackled then gains recorded in controlling/treating illnesses like HIV, TB, and malaria would be in dire situation [4].

1.1 Antimicrobial resistance

The case of methicillin-resistant *Staphylococcus aureus* was reported in the 1950s [8]. *Salmonella*, *Shigella*, and *Escherichia coli* were intestinal bacteria that were resistant to various antimicrobial treatments in the late 1950s and early 1960s. These resistant strains led to significant life, financial, and clinical loss, primarily in developing countries. However, it was seen as a minor health issue limited to intestinal bacteria in the developed world. This false belief was dispelled in the 1970s when it was discovered that *Haemophilus influenzae* and *Neisseria gonorrhoeae* are ampicillin-resistant, with *Haemophilus* also being reported to be resistant to tetracycline and chloramphenicol [5]. Antibiotics destroy delicate germs, but they leave behind resistant pathogens, which eventually proliferate and flourish due to natural selection [9].

AMR is the main issue when treating a certain disease for a long time. AMR occurs when microorganisms stop responding to medications that once killed them. The rise in AMR is caused by both microbial behavior and how people take antimicrobial medications. This resistance could be extremely harmful since it might make some illnesses impossible to cure, which could result in serious consequences or even death. To combat AMR, researchers are striving to create novel therapeutics [9].

Bacteria can develop antibiotic resistance in addition to the intrinsic mechanism of resistance. Other mechanisms may also contribute to the development of antibiotic resistance in bacteria. These include antibiotic efflux or poor drug penetration that lowers the antibiotic's intracellular concentration, modification of the antibiotic's target site caused by posttranslational target modification or genetic target mutation, and inactivation of the antibiotic through modification or hydrolysis [5].

1.1.1 Virulence

Human skin, mucous membranes, and internal organs all contain bacteria. The ability of bacteria to cause disease is known as pathogenicity, and a pathogen carries a number of elements known as virulence that enable the bacterium to enhance its level of pathogenicity. Toxicity and invasiveness are two of a pathogen's most crucial characteristics that aid in the development of a disease. Both virulence and the state of the host's immune system have the potential to affect the final balance of a bacterial illness course. The coevolution of the host and bacterium may have occurred over a period of millions of years. During this time, pathogens have altered their virulence to adapt to the host's immune system [10].

1.2 Global concern of AMR

AMR is an increasing challenge to the global economy, human health, and sustainable development due to the indisposition, death, and economic cost it causes [7, 11]. AMR is a global challenge; however, antibiotics have contributed immensely to medicine by treating bacterial infections in animals [5]. Inappropriate use of antibiotics is widely regarded to contribute to the growing challenge of resistant bacteria [12]. Sir Alexander Fleming, who discovered antibiotics, had cautioned that the public would demand antibiotics and this would begin an era of abuses. Therefore, the primary driver of resistance development is the overuse of antibiotics [9].

The infections that are caused by multidrug-resistant pathogens are linked with a high rate of mortality [13, 14]. The coronavirus pandemic has led to an increase in multidrug-resistant pathogens and this heightens the challenge of AMR [15]. The resistance of antibiotics is associated with: pumping antibiotics out of cell, decreasing uptake of antibiotics, distorting the target to reduce the binding of antibiotics with the target, and enzyme inactivation [16].

By 2050, AMR-related mortality is estimated to outpace diseases like cancer and diabetes, killing 10 million people annually [17, 18]. Major concerns about the security of the global health system include the escalation of AMR and the dearth of novel medications to treat drug-resistant bacterial illnesses [7]. The World Economic Forum, the World Health Organization (WHO), the Centre for Disease Control and Prevention (CDC), and the Infectious Diseases Society of America have all identified antibiotic resistance as a global public health concern [19, 20]. The World Health Assembly asked WHO to submit a worldwide action plan to address the issue of antibiotic resistance [21].

People in the UK have voted for a government-sponsored $10 million award (longitude award challenge) to find new ways to prevent antibiotic resistance [22, 23]. According to the US President's Council of Advisors on Science and Technology's recommendations, President Barack Obama in the United States instructed the National Security Council to create a comprehensive national action plan (NAP) to combat antibiotic resistance by 2015 [24, 25]. Multidrug-resistant (MDR) bacterial infections have been linked to over 8 million hospital stays and are currently costing healthcare systems over $20 billion, according to research by the WHO [26].

AMR is a critical issue that needs to be addressed on a global scale since it has a detrimental influence on patient safety, public health, and clinical outcomes. It also poses a threat to the advancements made in contemporary medicine, as it makes it difficult to treat severe infectious diseases effectively given how infrequently new antibiotics are being developed [27]. This places a significant load on healthcare systems all across the world [28]. Many nations have created and are putting into practice their NAPs for AMR [29].

1.3 Causes of AMR

Microbial resistance results because of improper antibiotic usage [30]. The overuse of antibiotics is discouraged nevertheless overprescription is still practiced around the world [31]. The misuse of antibiotics is facilitated by self-medication behaviors and the unrestricted procurement of antibiotics without prescriptions, particularly in low-income nations [31, 32]. It is pertinent to control AMR because of its potential to incur US $1 trillion in healthcare costs by 2050 if left unchecked [33, 34]. The improper use of antibiotics is present in hospitalized patients and community [35, 36]. About 62% of antibiotics sold in community pharmacies around the world are given out with no prescription [37]. Community pharmacists frequently give antibiotics without a prescription due to a variety of factors including complacency, patient pressure, and fear of losing customers [38, 39]. There are various studies that have confirmed inappropriate prescription of antibiotics [40, 41]. A meta-analysis revealed that in a particular community pharmacy setting, 62% of antibiotics were dispensed without a prescription to treat illnesses that do not even require antibiotic therapy [42].

Furthermore, infection control guidelines, sanitation settings, water hygienic practices, diagnostic and treatment procedures, drug quality, and travel or migration quarantine are additional significant factors that have the potential to cause antibiotic resistance. The exchange of genetic material between organisms enhances the spread of antibiotic resistance, in addition to the mutation of numerous genes located on the chromosome of the microbe [5]. Resistance results from diverse genetic mutations and alterations that make microorganisms less sensitive to certain types of antibiotics. As a result, infections become more difficult to treat and rates of transmission, sickness severity, and mortality considerably rise [43, 44]. Antibiotic resistance has resulted because bacterial infections are becoming more challenging to treat because of pharmaceutical industry's failure to produce new medications [40].

The lack of antimicrobials has made AMR a serious global problem. The cost of medications, the length of hospital stays, and the overall cost of healthcare services all rise as a result of AMR [45]. AMR is associated with poverty and the WHO has posited that by 2050 about 28 million could be pushed into extreme poverty [46]. Inappropriate management of antimicrobial agents has been shown to have a substantial impact both in developed and developing nations worldwide. The population's general lack of awareness and perhaps the prescribers' lack of knowledge are the main factors in the misuse of antibiotics. This may cause them to select the wrong treatment course or therapeutic agent, increasing the likelihood of the emergence of microbial resistance [47, 48].

1.4 Solution to AMR

Despite the WHO's repeated warnings new antibiotics are needed to address the growing threat of antibiotic resistance [49]. The rise of AMR rates has prompted calls from the WHO and UN urging nations to create NAPs [50]. Otherwise, AMR rates will increase significantly, having a negative influence on death, morbidity, and expenditures [51, 52].

Five strategic goals are outlined in the Global Action Plan on AMR as a guide for nations creating their NAPs on AMR:

Objective 1: Improve public knowledge and comprehension of AMR through effective outreach, instruction, and training.
Objective 2: Through observation and research, strengthen the body of information and evidence.
Objective 3: Reduce the likelihood of infection by implementing efficient sanitation, hygiene, and infection prevention practices.
Objective 4: Optimize the use of antimicrobial drugs for both human and animal health.
Objective 5: Develop the financial justification for long-term investments that consider the requirements of all nations and boost spending on innovative drugs, diagnostic equipment, vaccinations, and other interventions [53].

Utilizing antimicrobial drugs sparingly and for the shortest possible time is a crucial step in preventing and reducing resistance [54]. To promote appropriate antibiotic use in community settings, approaches like encouraging prescribers to include the diagnosis on the prescription, implementing delayed prescribing, point-of-care testing, and creating collaborative practice agreements can be effective options [55, 56]. A global action plan on antibiotic resistance is necessary in order to enrich public perceptions of antibiotic resistance, lowering the frequency of infections, stepping up monitoring and research efforts, and maximizing the use of antibiotics [57]. To ensure the best possible use of antibiotics in the public, proper dispensing procedures at community pharmacies are crucial in this situation [58]. To clarify, analyze, and examine pharmacists' and patients' perspectives, beliefs, and feelings, qualitative research methods are useful in pharmacy studies [59].

A distinct and fruitful strategy to combat AMR could be the use of alternative medicines for the treatment and control of infectious diseases. These treatments include vaccinations (against MRSA and MDR *Mycobacterium tuberculosis*), biological therapy (application of monoclonal antibodies, insulin, erythropoietin, etc.), and anti-virulence techniques (to manipulate the virulence components of bacteria) [60, 61]. Additionally, herbal remedies have illusive properties; yet, there is a strong argument that they might be a workable alternative [5].

2. Antimicrobial stewardship

One of the methods for reducing antibiotic resistance is known as antimicrobial stewardship (AMS) [62]. It ensures that the best antimicrobial therapies are chosen, administered, and continued for the shortest possible time to produce the best therapeutic results with the least amount of chance that the patient may experience adverse effects or the emergence of AMR [63]. Succinctly, AMS programs work to raise the success rates of treating infections, decrease treatment failures, and correctly prescribe therapy and prophylaxis [64]. The WHO has pushed for and taken the lead in developing frameworks for antimicrobial stewardship programs (ASPs) by working with other professional, national, and international organizations [53, 65]. AMS is accomplished by preventing the unnecessary use of medications and by offering focused, targeted care when it is appropriate in order to raise the standard of patient care [66].

Antibiotic stewardship is a method that makes sure antibiotics are used correctly and in sufficient amounts when necessary. Additionally, this assures that antibiotic use yields the most advantages, stops the spread of infections, and improves both health and economic benefits [67]. Evidence demonstrates that antibiotic stewardship initiatives are successful in encouraging judicious antibiotic usage and enhancing clinical results in the hospital context [68]. Reviews have demonstrated the effectiveness of ASPs in boosting antibiotic policy adherence; reducing the length of antibiotic therapy; decreasing antibiotic resistance, morbidities, mortalities, healthcare-associated infections, costs, and extended hospital stays [69]. ASPs help to promote achieving the sustainable development goals (SDGs) [70]. A thorough stewardship program should include constant monitoring and audit of antibiotic consumption patterns with clinician feedback; the development of an antibiotic formulary; nonstop educational initiatives; antibiotic de-escalation to ease the switch from broad spectrum to narrow spectrum antibiotics; the development of infection treatment guidelines; antibiotic restriction programs; and the enforcement of proper medication regimens [71].

Medical professionals, clinical microbiologists, pharmacists, nurses, and/or administrative staff typically make up multidisciplinary teams for ASPs and the interventions they use can vary greatly depending on the healthcare system and cultural context [72]. ASPs must be implemented in order to improve the prudent use of antibiotics in healthcare facilities and the general public [73, 74].

2.1 Objectives of AMS

Reduced antibiotic abuse, lower healthcare costs, improved clinical outcomes, and a decrease in AMR are the objectives of AMS [62]. The ASPs are targeted to ensure the responsible use and continuous efficacy of available antimicrobials through a wide range of treatments incorporating quality improvement efforts [4].

2.2 Role of pharmacists in AMS

According to the CDC and the American Society of Health-System Pharmacists (ASHP), pharmacists have a significant role to play in infection prevention and control programs which are important to AMS [75, 76]. Pharmacists contribute to AMS programs in the following ways: leading of AMS programs in healthcare facilities, formulary development; automatic change and stop orders; dose optimization therapeutic drug monitoring; and providing empirical antimicrobial guidance [62]. Additionally, AMS activities led by pharmacists maximize patient clinical outcomes, enhance the proper administration of antibiotics, and lower the cost of antibiotics [68, 77, 78].

Lack of training and information are among the major obstacles that affect the participation of pharmacists in AMS [62, 79]. Patient education relies heavily on pharmacists and numerous studies have demonstrated the benefits of instructing patients on proper antibiotic usage [70]. Therefore, it is advised that undergraduate pharmacy students should be trained in the areas of antimicrobial therapy, AMR, and AMS [62]. Pharmacists play critical roles in ASPs like: taking part in antimicrobial ward rounds; editing and writing antimicrobial guidelines; keeping track of antimicrobial consumption and spending; and educating healthcare professionals about antimicrobials [79]. Community pharmacies are the best place to introduce ASPs because of the effect antibiotic resistance has on the patient's condition and the cost to society [80]. Community pharmacists play a crucial role in patient management since they are frequently the first healthcare providers that people see about their condition [81]. They are crucial stewards of antibiotics and play a significant role in informing, educating, and counseling patients and healthcare professionals on how to use antibiotics properly [82].

In Sub-Saharan Africa (SSA), there is a shortage of infectious disease pharmacists. However recent studies have demonstrated that with the right training, pharmacists in SSA can create and oversee AMS programs that are comparable to those in the UK or the USA [4]. Existing research indicates that pharmacists are crucial in developing and managing antimicrobial guidelines and policies; evaluating individual patient regimens to optimize therapy; auditing antimicrobial consumption outcomes both prospectively and retrospectively; and training clinical teams and patients [83]. Additionally, it has been noted that doctors frequently rely on the advice and knowledge of pharmacists about antibiotic stewardship [84]. Studies have shown that AMS with a pharmacist reduced antibiotic consumption compared [85].

2.3 Implementation of AMS across the globe

Due to the growing concern about AMR and the drying up of the supply of new antibiotics, there has been a global push for the implementation of ASPs [72]. More detailed instructions on how to set up, carry out, and assess efficient AMS programs at the national and healthcare facility level are increasingly needed, particularly in developing countries [53]. The adoption of AMS principles in the real world is thought to depend heavily on education [86]. Healthcare systems around the world have embraced AMS as a crucial strategy for combating AMR [87].

The implementation of AMS in Nigerian hospitals is insufficient even though AMS is a part of the Nigerian National Antimicrobial Resistance Action Plan 2017–2022 [79]. The enforcement of a law banning the distribution of antibiotics without a prescription among community pharmacists in Saudi Arabia dramatically reduced such malfeasance [88]. In 2007, the Thailand Ministry of Public Health started a campaign "Antibiotic Smart Use" to encourage prudent antibiotic use in Thailand hospitals. In 2016, the Royal Thai Government released a National Strategic Plan on AMR [64].

According to a study conducted in Australia, clinical leaders' involvement and executive-level support were essential for the successful implementation of AMS programs [89]. In order to implement a comprehensive and sustainable ASP it is pertinent to take into consideration the following: the need for developing formal AMS policies and incorporating them into the clinical governance structure; establishing a multidisciplinary approach to AMS with clear roles and responsibilities for each team member and importance of organizational investment in staffing, information management systems, and staff education [64].

3. Plants contribution to medicine

Traditional medicine is acknowledged by the WHO as an essential alternative healthcare delivery system for the majority of the global population [90]. According to WHO, more than 80% of people in developing nations still rely on herbal medicine to treat diseases [91]. Globally, herbal medicine either forms the backbone of healthcare delivery or works in tandem with it [92]. Numerous plants have secondary metabolites also referred to as bioactive metabolites because of their antimicrobial properties [93].

The large and diverse groups of organic compounds that make up the secondary metabolites in plants are produced in small amounts, and they play no direct role in vital processes like photosynthesis and respiration [94]. The primary function of a plant's secondary metabolites is to defend itself against pathogens and predators. These secondary metabolites have significant applications in modern medicine as they are useful for treating or preventing some human ailments [95]. Examples of secondary metabolites include: tannins, flavonoids, terpenoids, and saponins and they possess antibacterial properties [90].

The combination of compounds known as secondary metabolites found in plants is primarily responsible for the good therapeutic effects of plant materials [96]. Secondary metabolites are efficient and more affordable than conventional drugs. They have been seen to be the preferred first-line therapy option for people, particularly traditional healers, in treating infections and other disorders [97]. About 25% of drugs produced are derived from plants and even synthetic drugs derive their structures from natural products. This is because plant secondary metabolites provide protection against microbial infection and insects [2].

Plant defense mechanisms vary depending on their unique needs and are influenced by physiological conditions, climatic changes, and environmental factors [98, 99]. Secondary metabolites can treat diseases by themselves or in conjunction with other substances or metabolites. Numerous studies have shown that such combinations can improve the effectiveness of a disease's treatment [100]. Due to their bioactivity, secondary metabolites have historically been employed in medical systems across the globe [101].

Natural products derived from medicinal plants give room for the discovery of new drugs [102]. There are many medications made of various secondary metabolites that have the potential to address the issue of drug resistance and open up a new avenue for researchers to find new medications [101]. Concern over antibiotic resistance is now widespread. The rise of MDR bacteria poses a danger to the therapeutic efficacy of several currently available medications. Treatment relies heavily on the use of plant extracts and phytochemicals, both of which have well-known antibacterial effects [103]. In order to survive and thrive in the natural environment, plants produce a variety of biologically active substances that shield them from pests, abiotic stress and disease infections [104]. In developed and developing nations, there is increased interest in medicinal plants due to their potential to cure infectious diseases [105].

3.1 Pharmacological activities of secondary metabolites

Secondary metabolites perform biological activity that neutralizes bacteria or animal cells by interacting with specific targets within those structures. However, the variety of metabolic routes that plants take to produce these metabolites ensures the presence of unique structures in these defense molecules that can be used to create new medications and pharmaceuticals. Because of this, plants are a significant source of compounds that can be utilized to enhance health and/or treat ailments [94]. Secondary metabolite chemicals found in many plants such as phenolics, alkaloids, carotenoids, and anthocyanins that accumulate in vegetables and fruit, may act as antioxidants. Antioxidants have the potential to lessen the oxidative and structural harm brought on by free radical molecules, and they are crucial in the prevention of cancer, the slowing of aging, UV protection, and the reduction of tissue inflammation in maintaining human health [105].

The antibacterial action of polyphenols is based on their capacity to prevent the development, reproduction, respiration, and any other essential function of microbes. This effect is caused by the oxidation of particular enzymes, which also inhibit several vital processes like respiration. Additionally, it has been claimed that polyphenols inhibit the creation of proteins in bacteria by binding to DNA chains. According to some scientists, certain polyphenols may be able to rupture the cell walls of microorganisms, leading to cell apoptosis. Due to their lipophilic character, monoterpenes are also known to interact with the phospholipids found in the cell membranes of numerous bacteria [94].

The most common characteristic for bacteria to live in an unfavorable environment are bacterial biofilms. Due to their high antibiotic tolerance, biofilms are one of the causes of chronic, nosocomial, and medical device-related infections, which present a significant challenge to the healthcare system. Several plant-derived substances, including phenylpropanoids, terpenoids, betulinic and ursolic acids, and alkaloids, such as berberine, indole, and chelerythrine, exhibited anti-biofilm activity toward *Pseudomonas aeruginosa*, *Klebsiella pneumoniae* and *staphylococcus* biofilms. The disruption of intercellular communication, disturbance in cell-to-cell coaggregation, inhibition of cell mobility, inactivation of bacterial adhesins, or stimulation of bacterial dispersal are the suggested mechanisms of plants' secondary metabolites in inhibiting bacterial biofilm [106]. The screening of plant extracts and natural products for antimicrobial activity has revealed that medicinal plants are a potential source of novel anti-infective medicines [107].

3.2 Mechanism of action of secondary metabolites

Secondary metabolites can affect the microbial cell in a variety of ways, including disruption of the cytoplasmic membrane's efflux system, interaction with membrane proteins, impacting DNA/RNA synthesis and function, impairing enzyme synthesis, causing cytoplasmic constituents to coagulate, and interfering with normal cell communication (quorum sensing) [106]. Terpenoids, alkaloids, and flavonoids are some of the substances that are currently utilized as medications or dietary supplements to treat or prevent a variety of illnesses like cancer. According to estimates, 14–28% of higher plant species are used medicinally, and research into ethno-medical plant use led to the discovery of 74% of pharmacologically active plant-derived components [108].

As a result, we are aware that the alkaloids have the capacity to intercalate with DNA, interrupting transcription and replication, and that they can also suppress cell division, leading to cell death. For instance, when *Streptococcus agalactiae* interacts with berberine it can severely disrupt the structure of bacterial cell membranes and prevent the creation of proteins and DNA. The action of flavonoids on the membrane of the microbial cell is what gives them their antibacterial properties; they interact with membrane proteins found on bacterial cell walls, making the membrane more permeable and disrupting it. Terpenes possess the capacity to damage microbial membranes and this is largely responsible for their antimicrobial effects [96].

3.3 Potential of natural products for combating AMR

The chemistry of natural drug products versus synthetic pharmaceuticals differs greatly, with natural products having a wider diversity of chemicals. The nitrogen, phosphorus, sulfur, and halogen content of natural products is lower, while their structural complexity, scaffold variation, stereochemistry, ring system diversity, and carbohydrate content are increased [106]. Compared to current antibiotics, crude extracts from medicinal plants have proven to be more therapeutically efficacious and less harmful. The defense against free radicals and pathogenic bacteria is greatly aided by phytochemical substances, particularly flavonoids and other natural compounds. In order to tackle infectious disorders linked to drug-resistant microbes and oxidative stress, there is a compelling need to investigate new and more effective antimicrobial/antioxidant substances of natural origin [107].

Natural plant extracts have a lower likelihood of developing resistance since they include numerous and complex phytochemicals [109]. A single plant contains a variety of phytochemicals so plant extracts can fight infections in a variety of ways. *Thymus vulgaris* essential oil has been shown to be effective against *Haemonchus contortus* that is resistant to benzimidazoles, macrocyclic lactones and imidazothiazoles. It can target different stages of the parasite's life cycle, including the ability to inhibit egg hatching, larvae motility, and development [110]. Natural plant extracts and compounds can be used with conventional antibiotics to increase the antibacterial efficacy in addition to having multiple modes of action. It was discovered that the alkaloid berberine, which is found in many plants, works in conjunction with fluconazole to combat *Candida albicans* that are resistant to the drug. Fluconazole may raise the intracellular content of berberine by rupturing the fungal cell membrane and this would enhance the effect of berberine [111]. Disruptions of cell membrane structures and function are mechanisms that some plant bioactive compounds like geraniol, monoterpene linalool, cinnamaldehyde, eugenol, and carvacrol, show their antifungal activity [106].

4. Future perspectives

Some antimicrobial metals kill MDR bacteria and selectively disrupt metabolic pathways [112]. Metal nanoparticles have the potential to threaten bacterial survival and antimicrobials containing silver can impact physical stress on cells of bacteria [113]. Also, gallium can interfere with the metabolic pathways of bacteria [114]. Genetically engineered bacteria can be used to target pathogens. A good example is the use of *E. coli* to secrete antimicrobial peptides in response to quorum-sensing molecules released by *P. aeruginosa* [113]. Furthermore, synthetic drugs can be combined with natural products to tackle AMR. The combination of fluconazole and berberine may be an effective combination to boost fluconazole's effectiveness in fluconazole-resistant *C. tropicalis* [106].

5. Conclusion

Antibiotics have played a crucial role in medicine by reducing morbidity and mortality rates. Without antibiotics, some successful aspects of modern medicine may not have been achieved. However, the inappropriate use of antibiotics has resulted in to the development of AMR. This poses a threat to human health, global economy, and sustainable development. World Economic Forum and WHO have labeled AMR as a matter of global public concern and many nations have taken stance on tackling this menace. This has led to increased efforts to tackle the challenge of AMR. AMS has been identified as one of the methods of reducing AMR. The method is focused on preventing unnecessary use of antibiotics in order to prevent the occurrence of AMR. Some of the objectives of the ASPs are: reducing antibiotic use and lowering health-care costs. Multidisciplinary teams are used to champion the ASPs and pharmacists play crucial roles in these teams. Another way to tackle AMR is the development of new antibiotics. Plants have contributed immensely to healthcare delivery across the globe since prehistoric times. Bioactive secondary metabolites have pharmacological functions for treating diseases and they possess the potential to contribute to tackling AMR. The study also highlights future perspectives that should be considered in combating AMR.

Author details

Khalifa Musa Muhammad* and Mansurat Oluwatoyin Shoge
Department of Chemistry, Air Force Institute of Technology, Kaduna, Nigeria

*Address all correspondence to: mkhalifam14@gmail.com

References

[1] Etebu E, Arikekpar I. Antibiotics: Classification and mechanisms of action with emphasis on molecular perspectives. International Journal of Applied Microbiology and Biotechnology Research. 2016;**4**:90-101

[2] Demain AR. Antibiotics: Natural products essential to human health. Medicinal Research Reviews. 2009;**29**(6):821-842. DOI: 10.1002/med.20154

[3] Nathan C, Cars O. Antibiotic resistance-problems, progress, and prospects. New England Journal of Medicine. 2014;**371**(19):1761-1763

[4] Otieno PA, Campbell S, Maley S, Arunga TO, Okumu MO. A systematic review of pharmacist-led antimicrobial stewardship programs in Sub-Saharan Africa. International Journal of Clinical Practice. 2022:3639943. DOI: 10.1155/2022/3639943

[5] Aslam B, Wang W, Arshad MI, Khurshid M, Muzammil S, Rasool MH, et al. Antibiotic resistance: A rundown of a global crisis. Infection and Drug Resistance. 2018;**11**:1645-1658

[6] Van Boeckel TP, Brower C, Gilbert M, Grenfell BT, Levin SA, Robinson TP, et al. Global trends in antimicrobial use in food animals. Proceedings of the National Academy of Sciences. 2015;**112**(18):5649-5654. DOI: 10.1073/pnas.1503141112

[7] World Health Organization. Antimicrobial Resistance: Global Report on Surveillance. Geneva, Switzerland: World Health Organization; 2014

[8] Saga T, Yamaguchi K. History of antimicrobial agents and resistant bacteria. Journal of the Japan Medical Association. 2008;**137**(3):513-517

[9] Babu PC, Meeravali SN, Azad S, Kumar KR. An overview on microbial infections and their antimicrobial resistance. Indo American Journal of Pharmaceutical Sciences. 2020;**7**(3):457-462

[10] Martinez JL, Baquero F. Interactions among strategies associated with bacterial infection: Pathogenicity, epidemicity, and antibiotic resistance. Clinical Microbiology Reviews. 2002;**15**(4):647-679

[11] Laxminarayan R, Duse A, Wattal C, Zaidi AKM, Wertheim HFL, Sumpradit N, et al. Antibiotic resistance—The need for global solutions. The Lancet Infectious Disease. 2013;**13**:1057-1098. DOI: 10.1016/s1473-3099(13)70318-9

[12] Bell BG, Schellevis F, Stobberingh E, Goossens H, Pringle M. A systematic review and meta-analysis of the effects of antibiotic consumption on antibiotic resistance. BMC Infectious Disease. 2014;**14**:1-25

[13] Abubakar U, Tangiisuran B, Elnaem MH, Sulaiman SA, Khan FU. Mortality and its predictors among hospitalized patients with infections due to extended spectrum beta-lactamase (ESBL) Enterobacteriaceae in Malaysia: A retrospective observational study. Future Journal of Pharmaceutical Sciences. 2022;**8**(1):1-8

[14] Abubakar U, Zulkarnain AI, Rodríguez-Baño J, Kamarudin N, Elrggal ME, Elnaem MH, et al. Treatments and predictors of mortality for carbapenem-resistant gram-negative bacilli infections in Malaysia: A retrospective cohort study. Tropical Medicine and Infectious

Disease. 2022;**7**(12):1-12. DOI: 10.3390/tropicalmed7120415

[15] Abubakar U, Al-Anazi M, Rodríguez-Baño J. Impact of COVID-19 pandemic on multidrug resistant gram positive and gram negative pathogens: A systematic review. Journal of Infection and Public Health. 2023;**16**(3):320-331

[16] Singh SB, Barrett JF. Empirical antibacterial drug discovery-foundation in natural products. Biochemical Pharmacology. 2006;**71**:1006-1015

[17] de Kraker MEA, Stewardson AJ, Harbarth S. Will 10 million people die a year due to antimicrobial resistance by 2050? PLOS Medicine. 2016;**13**(11):1-6. DOI: 10.1371/journal.pmed.1002184

[18] O'Neill J. Tackling drug-resistant infections globally: Final report and recommendations. The Review on Antimicrobial Resistance. 2016:1-84

[19] Michael CA, Dominey-Howes D, Labbate M. The antimicrobial resistance crisis: Causes, consequences, and management. Frontiers in Public Health. 2014;**2**:145

[20] Spellberg B, Srinivasan A, Chambers HF. New societal approaches to empowering antibiotic stewardship. Journal of the American Medical Association. 2016;**315**(12):1229-1230. DOI: 10.1001/jama.2016.1346

[21] Hoffman SJ, Caleo GM, Daulaire N, Elbe S, Matsoso P, Mossialos E, et al. Strategies for achieving global collective action on antimicrobial resistance. Bulletin of the World Health Organization. 2015;**93**(12):867-876

[22] Payne DJ, Miller LF, Findlay D, Anderson J, Marks L. Time for a change: Addressing R&D and commercialization challenges for antibacterials. Philosophical Transactions of the Royal Society B Biological Science. 2015;**370**(1670):1-12. DOI: 10.1098/rstb.2014.0086

[23] Luepke KH, Mohr JF. The antibiotic pipeline: Reviving research and development and speeding drugs to market. Expert Review of Anti Infective Therapy. 2017;**15**(5):425-433. DOI: 10.1080/14787210.2017.1308251

[24] Ventola CL. The antibiotic resistance crisis: Part 1: Causes and threats. Pharmacy and Therapeutics. 2015;**40**(4):277-283

[25] Landers T, Kavanagh KT. Is the presidential advisory council on combating antibiotic resistance missing opportunities? American Journal of Infection Control. 2016;**44**(11):1356-1359

[26] Nasr Z, Paravattil B, Wilby KJ. The impact of antimicrobial stewardship strategies on antibiotic appropriateness and prescribing behaviours in selected countries in the Middle East: A systematic review. Eastern Mediterranean Health Journal. 2017;**23**(6):430-440. DOI: 10.26719/2017.23.6.430

[27] Gillani SW, Shahwan MKS, Szollosi DE. A questionnaire based survey among pharmacy practitioners to evaluate the level of knowledge and confidence towards antimicrobial stewardship. Pharmacy Practice. 2023;**21**(1):1-9. DOI: 10.18549%2FPharmPract.2022.4.2757

[28] MacBrayne CE, Williams MC, Levek C, et al. Sustainability of handshake stewardship: Extending a hand is effective years later. Clinical Infectious Diseases. 2020;**70**(11):2325-2332. DOI: 10.1093/cid/ciz650

[29] Antimicrobial Resistance: A Manual for Developing National Action Plans. Geneva: World Health Organization; 2016

[30] Holmes AH, Moore LS, Sundsfjord A, Steinbakk M, Regmi S, Karkey A, et al. Understanding the mechanisms and drivers of antimicrobial resistance. The Lancet. 2016;**387**:176-187. DOI: 10.1016/s0140-6736(15)00473-0

[31] Roca I, Akova M, Baquero F, Carlet J, Cavaleri M, Coenen S, et al. The global threat of antimicrobial resistance: Science for intervention. New Microbes and New Infections. 2015;**6**:22-29. DOI: 10.1016/j.nmni.2015.02.007

[32] Al-Hamad A. Over-the-counter delivery of antibiotics: Are we sending the right message? American Journal of Infection Control. 2012;**40**:81-81

[33] Byrne MK, Miellet S, McGlinn A, Fish J, Meedya S, Reynolds N, et al. The drivers of antibiotic use and misuse: The development and investigation of a theory driven community measure. BMC Public Health. 2019;**19**:1-11. DOI: 10.1186/s12889-019-7796-8

[34] World Health Organization. Briefing to WHO Member States. 2023. Available from: https://apps.who.int/gb/MSPI/pdf_files/2023/03/Item1_22-03.pdf

[35] Abubakar U, Sulaiman SAS, Adesiyun AG. Utilization of surgical antibiotic prophylaxis for obstetrics and gynaecology surgeries in Northern Nigeria. International Journal of Clinical Pharmacy. 2018:**40**(5):1037-1043. DOI: 10.1007/s11096-018-0702-0

[36] Abubakar U, Amir O, Rodríguez-Baño J. Healthcare-associated infections in Africa: A systematic review and meta-analysis of point prevalence studies. Journal of Pharmaceutical Policy and Practice. 2022;**15**(1):1-12. DOI: 10.1186/s40545-022-00500-5

[37] Auta A, Hadi MA, Oga E, Adewuyi EO, Abdu-Aguye SN, Adeloye D, et al. Global access to antibiotics without prescription in community pharmacies: A systematic review and meta-analysis. Journal of Infection. 2019;**78**:8-18

[38] Zapata-Cachafeiro M, González-González C, Váquez-Lago JM, López-Vázquez P, López-Durán A, Smyth E, et al. Determinants of antibiotic dispensing without a medical prescription: A crosssectional study in the north of Spain. Journal of Antimicrobial Chemotherapy. 2014:**69**(11):3156-3160. DOI: 10.1093/jac/dku229

[39] Zawahir S, Lekamwasam S, Aslani P. A cross-sectional national survey of community pharmacy staff: Knowledge and antibiotic provision. PLoS One. 2019:**14**(4):1-15

[40] Abubakar U, Tangiisuran B. Knowledge and practices of community pharmacists towards non-prescription dispensing of antibiotics in Northern Nigeria. International Journal of Clinical Pharmacy. 2020;**42**(2):756-764. DOI: 10.1007/s11096-020-01019-y

[41] Abubakar U. Practices and perceptions of Nigerian community pharmacists toward antimicrobial stewardship program. International Journal of Pharmacy and Pharmaceutical Sciences. 2020;**12**(4):37-42

[42] Al-Shami HA, Abubakar U, Hussein MSE, Hussin HFA, Al-Shami SA. Awareness, practices and perceptions of community pharmacists towards antimicrobial resistance and antimicrobial stewardship in Libya: A cross-sectional study. Journal of

Pharmaceutical Policy and Practice. 2023;**16**(46):1-10

[43] Munita JM, Arias CA. Mechanisms of antibiotic resistance. Microbiology Spectrum. 2016;**4**:481-511. DOI: 10.1128/microbiolspec.VMBF-0016-2015

[44] World Health Organization. Antimicrobial Resistance. 2020. Available from: https://www.who.int/news-room/fact-sheets/detail/antimicrobial-resistance

[45] Ababneh MA, Nasser SA, Rababa'h AM. A systematic review of antimicrobial stewardship program implementation in Middle Eastern countries. International Journal of Infectious Diseases. 2021;**105**:746-752

[46] Antimicrobial Resistance and Primary Health Care. World Health Organization; 2018. Available from: https://apps.who.int/iris/bitstream/handle/10665/328084/WHO-HIS-SDS-2018.57-eng.pdf

[47] Pinder RJ, Berry D, Sallis A, et al. behaviour change and antibiotic prescribing in healthcare settings: Literature review and behavioural analysis. Public Health England. 2015. Available from: https://assets.publishing.service.gov.uk/government/uploads/system/uploads/attachment_data/file/774129/Behaviour_Change_for_Antibiotic_Prescribing_-_FINAL.pdf

[48] Yu B, Wang S, Yin X, Bai J, Gong Y, Lu Z. Factors associated with doctors' knowledge on antibiotic use in China. Scientific Reports. 2016;**6**(1):1-5. DOI: 10.1038/srep23429

[49] Woolhouse M, Waugh C, Perry MR, Nair H. Global disease burden due to antibiotic resistance state of the evidence. Journal of Globalization and Health. 2016;**6**(1):1-5

[50] Saleem Z, Hassali MA, Hashmi FK. Pakistan's National Action Plan for antimicrobial resistance: Translating ideas into reality. The Lancet Infectious Diseases. 2018;**18**:1066-1067

[51] Cosgrove SE. The relationship between antimicrobial resistance and patient outcomes: Mortality, length of hospital stay, and health care costs. Clinical Infectious Diseases. 2006;**42**:S82-S89

[52] Founou RC, Founou LL, Essack SY. Clinical and economic impact of antibiotic resistance in developing countries: A systematic review and meta-analysis. PLoS One. 2017;**12**:1-18

[53] World Health Organization. Antimicrobial Stewardship Programmes in Health-Care Facilities in Low-And Middle-Income Countries: A WHO Practical Toolkit. Geneva, Switzerland: World Health Organization; 2019

[54] Burgera M, Fouriea J, Lootsa D, Mnisia T, Schellacka N, Bezuidenhouta S, et al. Knowledge and perceptions of antimicrobial stewardship concepts among final year pharmacy students in pharmacy schools across South Africa. South African Journal of Infectious Diseases. 2016;**1**(1):1-7

[55] Dobson EL, Klepser ME, Pogue J, Labreche MJ, Adams AJ, Gauthier TP, et al. SIDP community pharmacy antimicrobial stewardship task force, outpatient antibiotic stewardship: Interventions and opportunities. Journal of the American Pharmacists Association. 2017;**57**:464-473. DOI: 10.1016/j.japh.2017.03.014

[56] Bishop C, Yacoob Z, Knobloch MJ, Safdar N. Community pharmacy interventions to improve antibiotic stewardship and implications for pharmacy education: A narrative

overview. Research in Social & Administrative Pharmacy. 2019;**15**:627-631. DOI: 10.1016/j.sapharm.2018.09.017

[57] Alkadhimi A, Dawood OT, Hassali MA. Dispensing of antibiotics in community pharmacy in Iraq: A qualitative study. Pharmacy Practice. 2020;**18**(4):1-9

[58] Torres NF, Solomon VP, Middleton LE. Pharmacists' practices for non-prescribed antibiotic dispensing in Mozambique. Pharmacy Practice. 2020;**18**(3):1-13. DOI: 10.18549/pharmpract.2020.3.1965

[59] Amin MEK, Nørgaard LS, Cavaco AM, Witry MJ, Hillman L, Cernasev A, et al. Establishing trustworthiness and authenticity in qualitative pharmacy research. Research in Social & Administrative Pharmacy. 2020;**16**(10):1472-1482. DOI: 10.1016/j.sapharm.2020.02.005

[60] Escaich S. Antivirulence as a new antibacterial approach for chemotherapy. Current Opinion in Chemical Biology. 2008;**12**(4):400-408

[61] Rex JH, Eisenstein BI, Alder J, et al. A comprehensive regulatory framework to address the unmet need for new antibacterial treatments. The Lancet Infectious Diseases. 2013;**13**(3):269-275. DOI: 10.1016/S1473-3099(12)70293-1

[62] Abubakar U, Sha'aban A, Mohammed M, Muhammad HT, Sulaiman SAY, Amir O. Knowledge and self-reported confidence in antimicrobial stewardship program me among final year pharmacy undergraduate students in Malaysia and Nigeria. Pharmacy Education. 2021;**21**(1):298-305

[63] Gerding DN. The search for good antimicrobial stewardship. The on Quality Improvement. 2001;**27**:403-404. DOI: 10.1016/S1070-3241(01)27034-5

[64] van Gulik N, Hutchinson A, Considine J, Driscoll A, Malathu K, Botti M. Barriers and facilitators to integrating antimicrobial stewardship into clinical governance and practice: A Thai case study. International Journal of Infection Control. 2020;**16**(2):1-11

[65] Barlam TF, Cosgrove SE, Abbo LM, et al. Implementing an antibiotic stewardship program: Guidelines by the infectious diseases society of America and the society for healthcare epidemiology of America. Clinical Infectious Diseases. 2016;**62**:e51-e77

[66] Charani E, Holmes AH. Antimicrobial stewardship programmes: The need for wider engagement. BMJ Quality and Safety. 2013;**22**(11):885-887

[67] Abushaheen MA, Muzaheed, Fatani AJ, Alosaimi M, Mansy W, George M, et al. Antimicrobial resistance, mechanisms and its clinical significance. Disease-a-Month. 2020;**66**:1-21

[68] Abubakar U, Syed Sulaiman SA, Adesiyun AG. Impact of pharmacist-led antibiotic stewardship interventions on compliance with surgical antibiotic prophylaxis in obstetric and gynecologic surgeries in Nigeria. PLoS One. 2019;**14**(3)

[69] Davey P, Brown E, Charani E, McNeil K, Brown E, Gould IM, et al. Interventions to improve antibiotic prescribing practices for hospital inpatients. Cochrane Database of Systematic Reviews. 2013;**30**(4):1-371. DOI: 10.1002/14651858.cd003543.pub4

[70] Drug-resistant Infections: A Threat to Our Economic Future. World Bank Report. Available from: https://documents1.worldbank.org/curated/en/323311493396993758/pdf/final-report.pdf

[71] Aryee A, Price N. Antimicrobial stewardship—Can we afford to do without it? British Journal of Clinical Pharmacology. 2015;**79**(2):173-181

[72] Chung GW, Wu JE, Yeo CL, Chan D, Hsu LY. Antimicrobial stewardship: A review of prospective audit and feedback systems and an objective evaluation of outcomes. Virulence. 2013;**4**(2):151-157

[73] MarkGilchrist M, Wade P, Ashiru-Oredope D, Howard P, Sneddon J, Whitney L, et al. Antimicrobial stewardship from policy to practice: Experiences from UK antimicrobial pharmacists. Infectious Disease Therapy. 2015;**4**:51-64

[74] Sanchez GV, Fleming-Dutra KE, Roberts RM, Hicks LA. Core elements of outpatient antibiotic stewardship. Mobidity and Mortality Weekly Report. 2016;**65**:1-12. Retrieved from: https://www.cdc.gov/mmwr/volumes/65/rr/rr6506a1.htm

[75] Ponto JA. ASHP statement on the pharmacist's role in antimicrobial stewardship and infection prevention and control. American Journal of Health-System Pharmacists. 2010;**67**:575-577. DOI: 10.2146/sp100001

[76] Centers for Disease Control and Prevention. Core Elements of Hospital Antibiotic Stewardship Programs. 2014. Available from: https://www.cdc.gov/antibiotic-use/healthcare/pdfs/core-elements.pdf

[77] Wang J, Dong M, Lu Y, Zhao X, Li X, Wen A. Impact of pharmacist interventions on rational prophylactic antibiotic use and cost saving in elective cesarean section. International Journal of Clinical Pharmacology and Therapeutics. 2015;**53**(8):1-11

[78] Brink AJ, Messina AP, Feldman C, Richards GA, Becker PJ, Goff DA, et al. Antimicrobial stewardship across 47 South African hospitals: An implementation study. The Lancet Infectious Diseases. 2016;**16**(9):1017-1025. DOI: 10.1016/S1473-3099(16)30012-3

[79] Abubakar U, Tangiisuran B. Nationwide survey of pharmacists' involvement in antimicrobial stewardship programs in Nigerian tertiary hospitals. Journal of Global Antimicrobial Resistance. 2020;**21**:148-153

[80] Paravattil B, Zolezzi M, Nasr Z, Benkhadra M, Alasmar M, Hussein S. et al. An interventional call-back service to improve appropriate use of antibiotics in community pharmacies. Antibiotics. 2021;**10**:1-8

[81] Saleem Z, Hassali MA, Hashmi FK, Godman B, Saleem F. Antimicrobial dispensing practices and determinants of antimicrobial resistance: A qualitative study among community pharmacists in Pakistan. Family Medicine and Community Health. 2019;7:1-9. DOI: 10.1136/fmch-2019-000138

[82] American Society of Health-System Pharmacists. ASHP statement on the pharmacist's role in primary care. American Journal of Health-System Pharmacists. 1999;**56**:1665-1667. DOI: 10.1093/ajhp/56.16.1665

[83] Liaskou M, Duggan C, Joynes R, Rosado H. Pharmacy's role in antimicrobial resistance and stewardship. Clinical Pharmacy. 2018;**10**

[84] Mohiuddin AK. Pharmacist-led antimicrobial stewardship. ACTA Scientific Medical Sciences. 2019;**3**(10):112-116

[85] Mahmood RK, Gillani SW, Saeed MW, Vippadapu P, Alzaabi MJMA.

Impact of pharmacist-led services on antimicrobial stewardship programs: A meta-analysis on clinical outcomes. Journal of Pharmaceutical Health Services Research. 2021;**12**(4):615-625

[86] Saleh D, Abu-Farha R, Mukattash TL, Barakat M, Alefishat E. Views of community pharmacists on antimicrobial resistance and antimicrobial stewardship in Jordan: A qualitative study. Antibiotics. 2021;**10**(4):1-11. DOI: 10.3390/antibiotics10040384

[87] Thomas B, Abdulrouf P, Elkassem W, Al Hail F, Nazar Z, Nasr Z, et al. Clinical and economic impact of antimicrobial stewardship interventions reported among hospital inpatients in the Middle East: A systematic review protocol. 2020. DOI: 10.21203/rs.3.rs-50679/v1

[88] Alrasheedy AA, Alsalloum MA, Almuqbil FA, Almuzaini MA, Aba Alkhayl BS, Albishri AS, et al. The impact of law enforcement on dispensing antibiotics without prescription: A multi-methods study from Saudi Arabia. Expert Review of Anti-Infective Therapy. 2020;**18**(1):87-97

[89] Loh JAM, Darby JD, Daffy JR, Moore CL, Battye MJ, Lorenzo YSP, et al. Implementation of an antimicrobial stewardship program in an Australian metropolitan private hospital: Lessons learned. Healthcare Infection. 2015;**20**:134-140. DOI: 10.1071/HI15015

[90] Mohammed H, Muhammad HL, Hussaini MA. Comparative secondary metabolite compositions and anti-microbial properties of n-hexane and ethyl-acetate fractions of Nelsonia campestris. GSC Biological and Pharmaceutical Sciences. 2019;**9**(1):046-052

[91] Adeyemi MM, Habila JD, Enemakwu TA, Okeniyi SO, Salihu L. Antimalarial activity of leaf extract, fractions and isolation of sterol from Alstonia boonei. Tropical Journal of Natural Product Research. 2019;**3**(7):221-224

[92] Okoro IO, Auguster O, Edith OA. Antioxidant and antimicrobial activities of polyphenols from ethnomedicinal plants of Nigeria. African Journal of Biotechnology. 2010;**9**:2989-2993

[93] Pandey A, Kumar S. Antibiotic activity of antimicrobial metabolites produced from soil microorganisms: An overview. International Journal of Pharmaceutical Research & Allied Sciences. 2015;**4**(4):28-32

[94] Mera IFG, Falconí DEG, Córdova VM. Secondary metabolites in plants: Main classes, phytochemical analysis and pharmacological activities. Revista Bionatura. 2019;**4**:1000-1009 DOI: 10.21931/RB/2019.04.04.11

[95] Roopan SM, Madhumitha G. Bioorganic Phase in Natural Food: An Overview. Cham, Switzerland: Springer; 2018. DOI: 10.1007/978-3-319-74210-6

[96] Gorlenko CL, Kiselev HY, Budanova EV, Zamyatnin AA Jr, Ikryannikova LN. Plant secondary metabolites in the battle of drugs and drug-resistant bacteria: New heroes or worse clones of antibiotics? Antibiotics. 2020;**9**:1-19. DOI: 10.3390/antibiotics9040170

[97] Manzo LM, Moussa I, Ikhiri K, Yu L. Toxicity studies of *Acacia nilotica* (L.): A review of the published scientific literature. Journal of Herbmed Pharmacology. 2019;**8**:163-172

[98] Ballhorn DJ, Kautz S, Heil M, Hegeman AD. Analyzing plant defenses in nature. Plant Signaling & Behavior. 2009;**4**(8):743-745

[99] Samuni-Blank M, Izhaki I, Dearing MD, Gerchman Y, Trabelcy B, Lotan A, et al. Intraspecific directed deterrence by the mustard oil bomb in a desert plant. Current Biology. 2012;**22**(13):1218-1220

[100] Wink M. Modes of action of herbal medicines and plant secondary metabolites. Medicine. 2015;**2**(3):251-286

[101] Kaushik B, Sharma J, Yadav K, Kumar PS, A. Phytochemical properties and pharmacological role of plants: Secondary metabolites. Biosciences Biotechnology Research Asia. 2021;**18**(1):23-35

[102] Hassan LG, Yusuf AJ, Muhammad N, Ogbiko C, Mustapha MD. In vitro phytochemical screening and anti-snake venom activity of the methanol leaf and stem bark extracts of *Leptadenia hastata* (Asclepiadaceae) against *Naja nigricollis*. Asian Pacific Journal of Health Sciences. 2020;**7**(3):11-14

[103] Jain C, Khatana S, Vijayvergia R. Bioactivity of secondary metabolites of various plants: A review. International Journal of Pharmaceutical Sciences and Research. 2019;**10**(2):494-504

[104] Tamasi AA, Shoge MO, Adegboyega TT, Chukwuma EC. Phytochemical analysis and in-vitro antimicrobial screening of the leaf extract of *Senna occidentalis* (Fabaceae). Asian Journal of Natural Product Chemistry. 2021;**19**(2):57-64

[105] Dhaniaputri R, Suwono H, Amin M, Lukiati B. Introduction to plant metabolism, secondary metabolites biosynthetic pathway, and in-silico molecular docking for determination of plant medicinal compounds: An overview. Advances in Biological Sciences Research, Volume 22. In: 7th International Conference on Biological Science (ICBS 2021). 2021. pp. 373-382

[106] Arip M, Selvaraja M, Mogana, Tan LF, Leong MY, Tan PL, et al. Review on plant-based management in combating antimicrobial resistance - Mechanistic perspective. Frontiers in Pharmacology. 2022. DOI: 10.3389/fphar.2022.879495

[107] Tsamo DLF, Tamokou JDD, Kengne IC, Ngnokam CDJ, Djamalladine MD, Voutquenne-Nazabadioko L, et al. Antimicrobial and antioxidant secondary metabolites from *Trifolium baccarinii* Chiov. (Fabaceae) and their mechanisms of antibacterial action. BioMed Research International. 2021;**3099428**:1-15

[108] Pagare S, Bhatia M, Tripathi N, Pagare S, Bansal YK. Secondary metabolites of plants and their role: Overview. Current Trends in Biotechnology and Pharmacy. 2015;**9**(3):293-304

[109] Gupta PD, Birdi TJ. Development of botanicals to combat antibiotic resistance. Journal of Ayurveda Integrative Medicine. 2017;**8**:266-275. DOI: 10.1016/j.jaim.2017.05.004

[110] Ferreira LE, Benincasa BI, Fachin AL, França SC, Contini SSHT, Chagas ACS, et al. *Thymus vulgaris* L. essential oil and its main component thymol: Anthelmintic effects against *Haemonchus contortus* from sheep. Veterinary Parasitology. 2016;**228**:70-76. DOI: 10.1016/j.vetpar.2016.08.011

[111] Li DD, Xu Y, Zhang DZ, Quan H, Mylonakis E, Hu DD, et al. Fluconazole assists berberine to kill fluconazole-resistant *Candida albicans*. Antimicrobial Agents and Chemotherapy. 2013;**57**:6016-6027. DOI: 10.1128/AAC.00499-13

[112] Lemire JA, Harrison JJ, Turner RJ. Antimicrobial activity of metals: Mechanisms, molecular targets and applications. Nature Reviews Microbiology. 2013;**11**:371-384

[113] Lobanovska M, Pilla G. Penicillin's discovery and antibiotic resistance: Lessons for the future? Yale Journal of Biology and Medicine. 2017;**90**:135-145

[114] Minandri F, Bonchi C, Frangipani E, Imperi F, Visca P. Promises and failures of gallium as an antibacterial agent. Future Microbiology. 2014;**9**(3):379-397

Chapter 4

Implementing Antimicrobial Stewardship in Various Healthcare Settings

Olanrewaju O. Popoola

Abstract

Antimicrobial resistance is a global health problem associated with mortality, morbidity, and socioeconomic losses. Given the rapid evolution of microorganisms and the scarcity of novel antimicrobials, it is important to implement strategies that promote appropriate, evidence-based use of antimicrobials to preserve their efficacy. These strategies and interventions are collectively referred to as antimicrobial stewardship (AMS). AMS interventions are focused on ensuring that the right antimicrobials are given at the right dose to the right patient, through the right route and for the right duration of time in order to improve patient outcomes, reduce side effects from inappropriate antimicrobial use and reduce the cost of therapy. AMS interventions include providing patient and public education; educating health professionals; ensuring evidence-based antimicrobial use; strengthening infection prevention and control practices; and prioritizing "one health" focused strategies to mitigate AMR in humans, animals, and the environment. Successful antimicrobial stewardship programs possess seven core elements as recommended by the CDC- leadership commitment, education of health professionals, accountability, drug expertise, action-oriented implementations, regular tracking of adherence to stewardship processes, and reporting. Stewardship interventions are successfully implemented when healthcare workers and stakeholders perceive the benefits of the program to be more than the effort required to implement the change.

Keywords: antimicrobial resistance, antimicrobial stewardship, interventions, one health, antibiotic resistance

1. Introduction

Antimicrobial resistance (AMR) is a global public health challenge that has continued to thrive, accounting for mortality, morbidity, and significant socioeconomic losses from prolonged illness, multiple therapies, and healthcare-associated infections. Microorganisms develop resistance to antimicrobials naturally in an effort to preserve their existence, however, antimicrobial misuse, poor infection control practices, and lack of adequate diagnostic tools have accelerated the emergence and spread of AMR [1]. AMR was the direct cause of 1.27 million deaths and played a role in 4.95 million deaths in 2019 [2]. If urgent interventions are not initiated, it could cause about $100 million in socioeconomic loss by 2050 [3].

Antibiotic stewardship (AMS) interventions are programs and strategies set up to reduce the emergence and incidence of AMR - related infections, improve antimicrobial use, and optimise patient outcomes. Antimicrobial stewardship programs ensure that the right antimicrobials are given at the right dose to the right patient, through the right route for the right duration of time [1]. These interventions are aimed at improving patient outcomes, reducing side effects from inappropriate antimicrobial use, and reducing the cost of therapy. Antimicrobial stewardship interventions should be implemented in all healthcare and non-healthcare settings including but not limited to hospitals (both in-patient and out-patient units), intensive care units, pharmacies, primary healthcare settings, nursing homes, veterinary hospitals, agricultural farms, animals houses, and in the community [1].

2. Antimicrobial stewardship interventions

To effectively implement antimicrobial stewardship interventions, you need to understand these interventions and how they can be implemented, here are some essential stewardship interventions.

2.1 Patient and public education

The general public is a major consumer of antimicrobials and should be educated on the appropriate use of antimicrobials, the administration, storage, and proper disposal of antimicrobials [4]. Awareness of AMR and its effect on morbidity, mortality, and socioeconomic impact should be explained. Information for the public should clearly address negative antimicrobial use practices observed over time and present proper alternative practices that promote proper antimicrobial use. Several studies have identified that people in many developing low- and middle-income countries (LMIC) stop using antimicrobials on feeling better, keep antibiotics at home for future use and use antibiotics based on recommendations from friends and family and this inappropriate use corresponds to the prevalence of AMR in such settings [5]. Educating the public about AMR and AMS can be done using broadcast media like television, radio, internet-based communication channels like certified websites dedicated to information on AMR and AMS, and social media. Additionally, print and multimedia materials can be made available to the public. All information materials should be presented in a simple, easy-to-understand manner, should be translated into the community's local language(s), and should be tailored to the local cultural context in which they are used. The public should be educated on inappropriate antibiotic use for viral infections like influenza, and the wrong use of antibiotics as growth promoters in agriculture. Patients and the general public should be educated on how AMR develops and spreads, as well as interventions to stop AMR [6]. Integration of the proper use of antimicrobials into high school curricula may promote awareness and change behaviour toward antimicrobial use at an early age [1].

Likewise, direct health professional to patient education should be encouraged. All health practitioners with direct access to patients using antibiotics such as doctors, nurses, pharmacists, primary care nurses, and infection control specialists should educate patients on proper antibiotic use, and debunk myths and false information on AMR [4].

2.2 Education of health professionals

Antimicrobial stewardship intervention requires a multidisciplinary approach involving doctors, pharmacists, nurses, infection control experts, data scientists, laboratory technicians and specialists, dentists, surgeons, veterinary doctors, and primary care doctors. They should be educated on proper antibiotic prescribing and dispensing, evidence-based prescription guidelines, antimicrobial surveillance, and the importance of the "one health" approach to address antimicrobial resistance [6]. Educational modules on AMR and AMS should be included in continuing professional development (CPD) programs. In-person educational sessions have been proven to have a greater and longer-lasting effect on prescribing behaviour than printed materials [1].

Locally-adapted presentations, workshops, simulations, and lectures on AMS should be encouraged within health facilities and institutions as part of CPD. Print materials and online learning on AMS should also be encouraged. Regional health authorities should provide timely information on recent and changing trends in AMR patterns to help health facilities tailor information sessions to represent the present reality. Education on infection prevention control practices should be made compulsory to limit the occurrence and spread of healthcare-associated infection and limit patient-to-staff disease transmission and vice versa [7]. Health authorities should promote multidisciplinary teamwork among all health practitioners to ensure effective AMR reporting, surveillance, and implementation of AMS programs.

2.3 Ensuring appropriate antimicrobial use

The inappropriate use of antimicrobials, especially antibiotics has accelerated antimicrobial resistance, therefore many antimicrobial stewardship interventions focus on limiting antimicrobial use and encouraging evidence-based use when necessary [4]. Most of the several strategies to improve antimicrobial use are focused on appropriate antimicrobial prescribing and dispensing, a few are discussed below.

2.3.1 Promoting adherence to regional and/or facility-specific antimicrobial prescribing guidelines

Information from antimicrobial stewardship surveillance reports should be used to develop region and facility-specific clinical guidelines. These guidelines are to provide evidence-based prescription practices. The adherence of physicians to prescription guidelines has been found to improve the quality of care [8]. Additionally, computerised decision support systems (CDSS) can help improve adherence to prescription guidelines and provide case-specific treatment plans when needed. Technology-integrated prescription guidelines are an improved effort toward evidence-based prescription of medicines and physicians should be encouraged to embrace and utilise such innovations [8].

2.3.2 Promoting evidence-based antimicrobial prescribing

Accurate laboratory testing provides evidence for diagnosis and treatment. When available, laboratory test results help identify disease-causing pathogens which aid diagnosis. This in turn ensures optimal antimicrobial use, limits the occurrence of drug toxicity, and prevents the occurrence of subsequent antimicrobial resistance [9]. Strengthening laboratory capacity to provide prompt diagnosis and antimicrobial susceptibility test results are strategies emphasised by the World Health Organisation (WHO) to improve accurate antimicrobial prescribing [9]. The flow of samples from the patient to the laboratory and the feedback of test results should be optimised to provide timely care. The laboratory staff should periodically provide a facility cumulative antibiotic resistance report, called an "antibiogram". This antibiogram will be used to update prescription guidelines and provide evidence-based antimicrobial treatment plan pending antimicrobial susceptibility results.

2.3.3 Reducing the prescription of broad-spectrum antibiotics

Broad-spectrum antibiotics (for example vancomycin, amikacin, and meropenem) have been proven to be effective against many species of both gram-positive and gram-negative organisms. The inappropriate use of these broad-spectrum antibiotics has been associated with the emergence of multidrug-resistant organisms [MDRO] [5, 10]. Therefore, to preserve the effectiveness of these broad-spectrum antibiotics, their prescription and use should be limited, and be determined by the approval of an infectious disease expert or consultant. Narrow-spectrum antibiotics that are only effective against a specific subset of organisms should be prescribed instead of broad-spectrum antibiotics.

2.3.4 Implementing delayed antimicrobial prescribing

This involves prescriptions issued by a doctor to a patient to be used at a later date if symptoms worsen. This intervention has been shown to reduce antibiotic use [11–13]. Within the period of delayed prescribing, non-pharmaceutical interventions can be recommended.

2.3.5 Restricting the use of antimicrobials

Restrictive antimicrobial use requires limiting the use of certain "reserve" antimicrobials (like carbapenems, vancomycin, piperacillin/tazobactam, and echinocandin antifungals) until authorised according to approved criteria. This is usually done to preserve the effectiveness of these broad-spectrum and last-line antimicrobials, and reduce the cost of therapy and potential toxicity associated with the use of these antimicrobials. These restricted antimicrobials can only be used when:

1. It has been authorised by an infection control specialist.

2. In certain indications/conditions.

3. For situations when the resistance profile of the organism warrants use.

4. When other antimicrobials are contraindicated or have failed

5. After discussion and approval with the antimicrobial stewardship team.

It is important that the use of these "reserve" antimicrobials are closely monitored, reviewed and reported. This strategy is usually implemented within a hospital, health facility, or at a regional level. An antimicrobial restriction policy and procedure document should be created to ensure compliance with restricting antimicrobial use, making it clear who is authorised to approve the use of the restricted antimicrobials and how approvals can be obtained [14, 15]. The WHO's AWaRe categorisation alongside hospital/regional antimicrobial surveillance reports can be used to formulate antimicrobial restrictive formulary [16, 17].

2.4 Strengthen infection control practices

Infection prevention and control (IPC) strategies can be implemented at health facilities and in the community. Infection prevention programs in health facilities are mostly focused on reducing healthcare-associated infections. At the core of infection prevention practices are rapid microbial diagnosis to quickly identify multidrug-resistant organisms, and rapid intervention to curb its spread. Infection control practices are optimised when healthcare providers are well educated on the importance of these practices and their compliance are regularly monitored [18].

Infection control activities in health facilities involve regular hand washing; use of protective, sterilised gloves, masks, eye protection devices (goggles), and other personal protective equipment (aprons and gowns) when interacting with patients; appropriate disinfection of medical equipment; regular environmental cleaning and disinfection; investigating potential disease outbreak and implementing control measures; identifying and vaccinating susceptible workers against vaccine preventative diseases; appropriate infectious waste management; and the installation of high-efficiency particulate absorbing (HEPA) filters into the building's construction to limit the spread of airborne infections.

The objective of Infection prevention and control at the community level is to minimise infection occurrence and spread between people. IPC strategies in the community focus on ensuring good hygiene and sanitation; promoting vaccination; encouraging the use of insect repellants; avoiding contact with confirmed cases of infectious diseases; practising safe sex; and ensuring appropriate use of medication [18].

2.5 Regulation of antimicrobial prescription, supply, and use

Suboptimal antimicrobial prescription practices have been linked with the emergence of resistant microorganisms, especially in LMICs [19]. Therefore, adherence to prescription guidelines and evidence-based antimicrobial prescribing based on laboratory analysis and data from antimicrobial resistance surveillance should be prioritised.

Pharmacists should ensure the continuous supply of antimicrobials and formulate a restricted antimicrobial formulary based on antimicrobial resistance patterns and the WHO's AWaRe classification. The shortage of essential narrow-spectrum antibiotics have contributed to the increased, suboptimal use of newer and broad-spectrum antibiotics potentially promoting an increase in antimicrobial resistance [20].

All health facilities should have access to continuous supply of antimicrobials. The antimicrobial stewardship pharmacists must ensure proper selection, procurement, and distribution of antimicrobials on the prescription formularies, including the reserved antimicrobials. Pharmacists should ensure that antimicrobials are of appropriate standard and properly stored to preserve efficacy. A critical antimicrobial formulary can be developed based on infectious disease epidemiology, antimicrobial resistance patterns, and drug availability. Tools for accurate antimicrobial use projections and inventory management should be employed to prevent shortages of these essential medicines [21]. Comprehensive procurement systems and robust supply chains should be engaged to ensure that antimicrobials are readily available and affordable [22]. Regulatory authorities should discourage pharmaceutical companies from advertising antimicrobials, especially broad-spectrum antibiotics to decrease patient demand for such medicines and prevent abusive use [23].

2.6 Implementation of a "one health" approach to limit AMR

The term "One health" refers to interdisciplinary, multi-sectoral initiatives to guarantee the best possible well-being for people, animals, and the environment. Antimicrobials are rightly used in humans and animals to treat diseases, however, the prolonged use of antimicrobials for prophylaxis in humans and animals; the mass administration of antimicrobials to animals (especially antibiotics that are critical to humans like third generation cephalosporins); and the unregulated use of antibiotics (e.g colistin) at low sub-therapeutic doses as growth promoters in agriculture, have accelerated the development of antimicrobial resistance – making animals reservoirs for antimicrobial resistant organisms and antimicrobial resistant genes. These resistant organisms and genes can easily be passed into the environment when excreted in urine and faeces [24]. Antimicrobials are used in aquaculture in feeds, given as injections, and sometimes directly applied to the water to prevent or treat diseases. When the water is released into the environment, it contaminates the environment with antimicrobials [25]. Wastewater from the production of antibiotics in pharmaceutical industries can contain traces of these antibiotics and are released directly into the environment [26]. Correspondingly, the improper disposal of expired antimicrobials (of which up to 90% of the active pharmaceutical ingredient may be present) also contributes to the direct deposits of antimicrobials into the environment [27]. Farmers have also been found to use antibiotics to preserve fruits from bacterial infections, leaving them with antibiotic residues.

When antimicrobials and their metabolites are in the environment (soil, water, or air), they are mostly present at sub-therapeutic concentrations, creating selective pressure on organisms around them, resulting in these organisms becoming resistant, making the environment a reservoir of antimicrobial resistant organisms and antibiotic resistant genes. Humans interact with animals and with the environment, this interdependence creates opportunities for antimicrobial resistant organisms and antimicrobial resistant genes to be transmitted within and between humans, animals, and the environment.

It is important that wastewater and solid waste contaminated with antibiotics or their metabolites are properly treated. Wastewater and sewage sludge can be treated to remove antimicrobials and pollutants chemically by coagulation and flocculation; by adsorption using activated charcoal; by physicochemical methods – electrochemical oxidation, ozone-based antibiotic degradation, and ultraviolet rays; using membrane technology like ultrafiltration, nanofiltration, and microfiltration; biological treatments [27], liming and composting [28].

2.6.1 One health focused strategies to limit AMR

1. Farmers, veterinarians, agriculturists, aquaculturists, and all stakeholders in the food industry need to be educated on antimicrobial resistance and how improper use of antibiotics contributes to it. They should be discouraged from using antimicrobials for prophylaxis unless recommended by a veterinarian or infection control expert. They should be encouraged to apply good agricultural principles, use clean

water, and observe good sanitation and hygiene to prevent infection in plants and animals. Veterinarians should be specially trained on antimicrobial stewardship practices, and encourage their clients to practice the same. Educative information on AMR may be disseminated through specific outreaches to farmers, publications in food, agro- and aqua industries, and professional development programs.

2. Countries should be encouraged to terminate the use of antimicrobials as growth promoters in line with the WHO's recommendations [6].

3. Regional and national integrative surveillance systems should be initiated and maintained to monitor AMR trends in animals, humans, and the environment.

4. Apply relevant infection control practices to limit infections and their transfer between and within humans and animals.

3. Antimicrobial stewardship programs (ASPs)

Antimicrobial Stewardship Programs (ASPs) are consciously coordinated programs that promote the appropriate use of antimicrobials [29]. The goals of ASPs [1] are to;

1. Improve the quality of patient care and thus increase positive patient outcomes.

2. Ensure optimal antimicrobial prescribing.

3. Decrease drug-related adverse drug reactions.

4. Decrease treatment failure and mortality.

5. Reduce the transmission rate of healthcare-associated infections (HAI).

6. Reduce the occurrence of and prevent antimicrobial resistance,

7. Ensure cost-effective therapy and reduce unnecessary health costs [21].

ASP can be implemented in all settings where antimicrobials are prescribed, dispensed, or used such as hospitals, nursing homes, intensive care units, pharmacies, and primary healthcare settings.

3.1 Core elements of antimicrobial stewardship programs

All ASPs must possess certain elements that are key to the success of such interventions. These elements are a broad category of strategies needed for the success of an ASP. The Centre for Disease Control (CDC) has highlighted seven core elements of successful antimicrobial stewardship programs [29].

3.1.1 Leadership commitment

The success of any ASP relies on the support of the leadership of the health facility, regulatory institutions, pharmaceutical industries, scientists, and health insurance.

Leaders support ASPs by providing adequate human resources, financial support, and information technology for the successful implementation of such programs. Executives of health organisations need to be well informed of the benefit of ASPs and their potential impact on patient outcomes. Leaders, managers, and board of directors can support ASPs by:

1. Publishing formal documents to support ASP interventions and identifying them as a top priority.

2. Incorporating stewardship-related activities in job descriptions of the antimicrobial stewardship team to afford enough time to oversee antimicrobial stewardship activities.

3. Ensuring enough time for staff from relevant departments to contribute to stewardship activities.

4. Include participation and adherence to antimicrobial stewardship guidelines or interventions in annual performance reviews.

5. Providing sustainable, sufficient, and budgeted financial support for stewardship activities such as salary, training cost and incentives, and information technology resources.

6. Support training and education of all healthcare workers involved in antimicrobial prescription and use.

7. Meet regularly with the AMS team to assess resources needed, and interventions required to overcome challenges that impede the success of ASPs [29].

Leadership at the regional and national level are also required to support ASPs to ensure multi-sectoral collaborations and formulation of regional or national action plans that align with the goals and activities of the ASP.

3.1.2 Accountability

ASPs require a leader or co-leaders, who are accountable for programs' management and outcomes. Co-leadership of a physician and pharmacists have been shown to be effective [29]. Clinical staff are better suited to lead ASPs than non-clinical staff. These leaders must possess effective leadership, management, and communication skills.

Leaders of ASPs should have regular stewardship rounds where they interact with physicians on their choice of antimicrobials, reasons for prescriptions, and pre-authorisation of the use of restricted antimicrobials. These rounds encourage a critical review and evaluation of antimicrobial choices and assessment of physician's adherence to prescription guidelines. Face-to-face interactions give physicians and the antimicrobial stewardship team a chance to exchange ideas in an open forum and offer real-time feedback when necessary [30]. These stewardship rounds provide an opportunity for learning and development, and can increase the confidence of physicians when prescribing antimicrobials especially when the antimicrobial stewardship team confirms their choice of antimicrobials [31], and reduce inappropriate use of broad-spectrum antimicrobials [32]. Leaders of ASPs should also provide reports of stewardship activities, interventions, and impact to the health facility and regional health leadership [33].

3.1.3 Drug expertise

Successful ASPs have a pharmacist leader or co-leader taking responsibility for appropriate antimicrobial use. Clinical pharmacists with training in antimicrobial stewardship and infectious diseases are more suitable and highly effective for this role [29]. After evaluating the patient's diagnostic results, pharmacists can offer clinical information to support decisions regarding the patient's antimicrobial medication regimen and length of treatment [34]. Additionally, they can help develop guidelines for the management of specific infections and particular organisms (such as multidrug-resistant *Staphylococcus aureus*) as well as educate health professionals on the use of evidence-based antimicrobial prescribing, and contribute to antimicrobial surveillance efforts [35].

3.1.4 Action

This involves Implementing recommended stewardship interventions like reducing the prescription of broad-spectrum antibiotics; restricting certain "reserve" antimicrobials to pre-authorisation from the antimicrobial stewardship team; and re-evaluating and assessing antimicrobial prescription 36–72 h after first administration- considering culture and antimicrobial sensitivity results, clinical scenario and patients response to therapy [33]. The reassessing of antibiotic prescription after 36–72 h after initiation is referred to as an "antibiotic timeout", and has been associated with reduced antimicrobial use, and improved patient outcomes [36–38]. During an antibiotic timeout, changes could be made to the antibiotic selected, duration of antibiotic use, route of administration such as the transmission of intravenous antibiotics to oral antibiotics, and/or dosage such as dose optimisation in patients with organ dysfunction. Pharmacists can also raise alerts when potential antibiotic related drug–drug interactions are observed (such as interactions between fluoroquinolones and metals like calcium, iron, and aluminium- and magnesium-containing antacids), monitor patient's drug levels, prevent unnecessary repetition of antibiotic prescriptions, and ensure antibiotics are not used beyond a reasonable time. Microbiological consultations are important for optimal antimicrobial selection and use, aseptic techniques should be employed to prevent contamination of laboratory cultures. Specific guidelines should be established to cater for infections caused by specific organisms like methicillin-resistant *S. aureus* (MRSA), multidrug-resistant *Clostridium difficile,* and specific conditions like community-acquired pneumonia [29]. Prospective audits of antimicrobial use which involves the review of antimicrobial use by an external expert is required periodically, to further improve antimicrobial use.

3.1.5 Tracking

This entails assessing the processes and outcomes of ASPs, such as adherence to prescription guidelines (process). And outcomes like a decrease in the use of broad-spectrum antibiotics, a drop in the number of antimicrobial days of therapy (DOT) per 1000 patient days, and an improvement in overall patient outcomes [29]. Additionally, antimicrobial use at a facility level, and the emergence of antimicrobial resistant diseases should be collated and monitored, and forwarded to regional and national health authorities [33]. Auditing of ASPs can help identify areas where more education or focused attention is needed, evaluate the effectiveness of AMS interventions and provide feedback on resistance patterns to regional authorities.

3.1.6 Reporting

Reports on antimicrobials prescribed, purchased, and dispensed should be made available to physicians, nurses, pharmacists, and relevant staff, along with specific action points. Information gotten from antimicrobial use audits should be shared with prescribers along with specific action points to drive compliance to guidelines. Antimicrobial resistance patterns should be evaluated and shared with the infectious disease and antimicrobial stewardship team to improve guidelines and drug formulary. Sharing reports can be used as a tool to increase compliance to prescribing guidelines and motivation to engage in antimicrobial stewardship activities. Facility-based data should also be made available to the hospital board, the ministry of health, and other regional or national health authorities to drive evidence-based interventions.

3.1.7 Education

Members of the healthcare team, including doctors, nurses, pharmacists, microbiologist, and laboratory staff should be well educated on the mechanisms and drivers of antimicrobial resistance; appropriate antimicrobial prescribing; individual roles in antimicrobial stewardship; benefits of antimicrobial stewardship; antimicrobial prescribing guidelines; antimicrobial pre-authorisation requirements; and the importance of antimicrobials stewardship intervention like antibiotic timeouts and antimicrobial use audits. Education can be done in formal and informal settings through lectures, posters, newsletters, case scenarios, educational video games, or electronic communications about antibiotic use with relevant departments [33, 39]. Education is most effective when initiated with interventions and monitored outcomes [33]. Patients should also be educated on appropriate antimicrobial use, the importance of antimicrobials, and potential side effects. Information to patients should be simple, adapted to the local language, and easy to understand.

3.2 Implementing antimicrobial stewardship programs

The first step in implementing an ASP is to form an antimicrobial stewardship committee. This committee should consist of a clinical doctor (preferably with training in infectious diseases and AMS, with keen interest in antimicrobial use and patient safety), a pharmacist (with training in infection prevention and control and antimicrobial stewardship), a nurse, a clinical microbiologist, infection control physician/expert, data analyst, information technology expert. Each member has unique roles and responsibilities.

3.2.1 Role of a physician in a stewardship team

Physicians are to develop treatment guidelines in collaboration with the infection control team and pharmacist, and also adhere to appropriate prescription practices and guidelines to ensure proper antimicrobial prescription and use. They are to ensure evidence-based decisions for infection management utilising data on antimicrobial susceptibility and laboratory diagnosis. When prescribing antimicrobials, physicians need to account for antimicrobial allergies, such as penicillin allergies, to limit adverse drug events. Physicians de-escalate the use of broad-spectrum and reserved antimicrobials to simple, narrow-spectrum regimens based on laboratory diagnosis, antimicrobial culture, and sensitivity. It is also important that they shorten the duration of antimicrobial therapy for surgical prophylaxis. In relating with patients, physicians can advise patients on appropriate antimicrobial use and how best to prevent infections. Senior physicians are to influence the prescribing of other physicians and facilitate peer-to-peer learning for improved adherence to established guidelines. Physicians are to take co-leadership of ASPs and communicate with facility leadership and stakeholders on the importance and impact of ASPs [40, 41].

3.2.2 Role of a pharmacist in a stewardship team

During antibiotic timeouts, pharmacists should assess and, if necessary, convert intravenous antibiotic regimens to oral regimens. As medicine experts, they are required to ensure the appropriate selection of antimicrobials based on laboratory test results and individualise dosing in accordance with clinical presentations, administration route, genetics, laboratory analysis, organ dysfunction, and comorbidities. Additionally, pharmacists are to collaborate with the infectious control team to develop treatment guidelines for specific infections (like community-acquired pneumonia) and specific organisms like vancomycin-resistant enterococci (VRE). Pharmacists participate in "antimicrobial timeouts" to assess the efficacy of antimicrobial prescribing practices, and "antimicrobial audits" to evaluate overall antimicrobial prescribing and dispensing practices. Based on antimicrobial susceptibility patterns, pharmacists are to create facility- and/or region-specific antibiogram to guide antimicrobial prescribing. They are also well skilled to formulate facility-based essential antimicrobial lists and restrictive formularies based on the WHO's AWaRe classification and regional antimicrobial resistance patterns [16]; intervene and prevent drug-drug and drug-food interactions; and ensure continuous availability of essential antimicrobials [34, 35].

3.2.3 Role of nurses in a stewardship team

Nurses must ensure the aseptic collection of laboratory specimens for laboratory investigations. Laboratory specimens must be handled with aseptic techniques to prevent contamination, which could result in the use of antimicrobials where there is no medical indication for their use. They can collaborate with the pharmacist to review laboratory results to help guide antimicrobial selection. They are to administer antimicrobials promptly and record antimicrobial use. Nurses monitor patients' progress and report signs of infection (including sepsis), allergies, and observed adverse drug events (like diarrhoea) to the physician and pharmacist. They serve as a vital communication channel between patients and doctors and assess patients' capacity to transition from intravenous drug therapy to oral therapy [42, 43].

3.2.4 Role of microbiologist in a stewardship team

To optimise the selection and use of antimicrobials, microbiologists conduct antimicrobial sensitivity tests and interpret laboratory test results. They evaluate novel diagnostic techniques and recommend their use in healthcare settings. Diagnostic tests that are affordable, sensitive, specific, and support rapid identification of organisms and possible resistance patterns should be prioritised. They review antibiograms based on antimicrobial surveillance reports, and report cases of antimicrobial resistance throughout the healthcare system. Lastly, they educate other members of the healthcare team on standard laboratory practices for optimal antimicrobial prescribing [44].

3.2.5 Infection control specialist

These specialists are to instil infection prevention and control measures that minimise infection risk. They promote the use of personal protective equipment. They develop and implement isolation guidelines and procedures for highly infectious and critical disease cases. And ensure aseptic techniques and strategies that prevent contamination of laboratory specimens, and limit exposure of the blood and body fluids of patients. Likewise, they ensure proper waste management, environmental hygiene, disinfection, and sterilisation as well as provide education and training on proper infection prevention and control strategies to healthcare professionals [18].

3.2.6 Role of data analyst in a stewardship team

Data provides evidence for effective decision-making and should not be trivialised. Data analysts collect, validate, clean, and analyse data on the antibiotic supply, prescription, and use. They inform health facility management, regional surveillance systems, and national surveillance systems of the findings of significant data analysis. They monitor and evaluate the outcomes of ASPs in line with predetermined indicators.

3.2.7 Role of information technology expert in a stewardship team

In broad terms, information technology experts essentially digitalise systems, structures, and processes for effective storage, timely communication and intervention. They integrate electronic records into the health system to facilitate efficient data collection, and rapid sharing of necessary information, like laboratory results, among health workers and the antimicrobial stewardship team. They maintain databases of antimicrobial resistance trends, and information on antimicrobial prescribing, purchase, and use. Similarly, they promote technology-aided drug prescribing and integrate computer provider order entry [CPOE] across the health facility for multidisciplinary, holistic patient care.

3.3 Strategic action plans for implementing antimicrobial stewardship programs

ASPs are implemented in all health settings where antimicrobials are prescribed, dispensed, and used. However, there is no one size fits all approach to implementing ASPs. Below are a few essential steps in implementing ASPs, these strategies should be adopted to fit the local context of the health setting in which they are applied.

3.3.1 Create an antimicrobial stewardship committee or team

The first step is to establish the antimicrobial stewardship committee or team. The team may stand alone or be integrated into the infection control or patient safety committee. For this team to be effective, it should have the support of the leadership of the health facility, major stakeholders, and managers in the health facility. The team is to be co-led by a physician and pharmacist with training in infectious disease control, and antimicrobial stewardship. When a physician is not available, a trained pharmacist can take up the leadership role.

3.3.2 Execute a facility-based SWOT analysis

The stewardship team is to carry out a SWOT analysis of the health system to evaluate areas of strength, weaknesses, opportunities, and threats. Already existing facility or regional prescription guidelines should be reviewed, and compliance with these guidelines be evaluated. Additionally, antimicrobial use data should also be evaluated. The focus of the SWOT analysis should be on assessing the availability of the core elements of AMS at the health facility. Resources - human and capital resources needed for the program should be assessed. In cases where a specialist member of the team like a pharmacist with training in infectious disease management is not available within a facility, then another similar expert should be identified off-site within another similar facility and can be consulted to collaborate and fill that gap [45].

3.3.3 Develop a focused action plan

A well developed and planned antimicrobial stewardship intervention is needed for the success of an ASP. All core elements of antimicrobial stewardship should be well established with support from the leadership of the health facility. Senior physicians should partner with the antimicrobial stewardship team to influence other physicians to implement antimicrobial prescription and use changes that come with the ASP.

Antimicrobial stewardship activities should be identified based on facility-specific peculiarities, after a SWOT analysis has been completed. Antimicrobial stewardship programs implemented in other similar facilities can be examined and adopted to fit the local context. Action plans should focus on initiating and maintaining behavioural changes to how antimicrobials are prescribed and used. The use of antimicrobial prescription guidelines, pre-authorisation of the use of restricted antimicrobials, limiting the use of broad-spectrum antibiotics, and evidence-based antimicrobial prescriptions backed up by laboratory results should be encouraged. Pharmacists should categorise antibiotics, and restrict access to their use based on the WHO's AWaRe classification, participate in antimicrobial use review (see **Table 1**) [17].

SMART (specific, measurable, attainable, realistic, and time-bound) goals for the stewardship program should be set, each member of the stewardship committee should understand their roles, responsibilities, and expected outcomes, and should be supported with the necessary authorisation needed to execute their task. Indicators to measure the success of the ASP should be well defined, and a period of time to evaluate the ASP based on the set indicator should be decided on whether every 3–6 months. Feedback from each evaluation process should be considered and necessary adjustments should be made to the antimicrobial stewardship plan.

3.3.4 Initiating sustained behavioural change in antimicrobial prescription practices

For a change process to be successful, all stakeholders involved must perceive the benefits of the change to be greater than the required efforts to actualise the change, the greater the perceived benefit, the quicker the adoption. The antimicrobial stewardship team must be able to effectively communicate the reason and benefits of changes in antimicrobial prescribing practices, and other stewardship activities to clinicians and stakeholders. Physicians must understand the proposed benefits of antimicrobial stewardship practices to the patients and also to them. The team must take into account factors that influence wrong prescribing practices and work on addressing them. It is important that all stakeholders are

WHO Classification	Antibiotics	
Access antibiotics – they possess narrow-spectrum antibiotic activity, they are affordable, safe with low resistance profile. They are usually effective against a wide range of commonly susceptible organisms. They should be used as first- or second-choice treatment option.	Amikacin	Cloxacillin
	Amoxicillin	Doxycycline
	Amoxicillin+ clavulanic acid	Gentamicin
		Metronidazole
	Ampicillin	Nitrofurantoin
	Benzathine	Procaine
	benzylpenicillin	benzylpenicillin
	Benzylpenicillin	Spectinomycin
	Cefalexin	Sulfamethoxazole+ trimethoprim
	Cefazolin	
	Chloramphenicol	
	Clindamycin	
Watch antibiotics – they are broader-spectrum antibiotics, and are usually more expensive than the access group. They should be used in clinically critical cases or cases where the causative organisms is likely to be resistant to the access group	Azithromycin	Ciprofloxacin
	Cef ixime	Clarithromycin
	Cefotaxime	Meropenem
	Ceftazidime	Piperacillin + tazobactam
	Ceftriaxone	
	Cefuroxime	Vancomycin
Reserve antibiotics – these are last-line antibiotics usually reserved to treat infections caused by multi drug resistant organisms	Ceftazidime + avibactam	Meropenem + vaborbactam
	Colistin	Plazomicin
	Fosfomycin	Polymyxin B
	Linezolid	

Table 1.
WHO's AWaRe classification of antibiotics.

consulted on the ASP and the changes that will be initiated. Having informal focus group discussions will help the stewardship team take into account the views of other health actors and limit potential resistance to stewardship programs when initiated and scaled up [45].

Behavioural changes and interventions initiated by the stewardship program will be speedily adopted and scaled up if they possess the seven CORRECT attributes.

1. Credible: Stewardship interventions should be evidence-based and should be supported by respected staff (like consultants and senior physicians) and leadership of the health facility.

2. Observable: There must be established evidence and testimonials that suggested interventions have yielded measurable success in similar settings. Study tours or site visits to such facilities with successful stewardship programs are encouraged so adopters see for themselves the benefits of such interventions.

3. Relevant: stewardship interventions must be adopted to solve established challenges in the local setting to which they are applied, making them relevant.

4. Relative advantage: The suggested interventions should present relative advantages over existing practices to the patient and health professional. Health facility leadership/management may further support interventions that manage human and financial resources better, and save cost. Additionally, the perceived benefit of the intervention must be greater than the perceived effort and time required to implement the intervention.

5. Easy: Interventions should be presented in an easy-to-understand and implement, simplified manner while preserving its essential components.

6. Compatibility: Stewardship interventions should align with the basic core values of the health setting in which they are initiated. Examine ways to implement stewardship interventions using already existing systems to limit excessive changes. Also, adopt stewardship interventions to the local context to address facility-based challenges.

7. Testable: Stewardship interventions should be testable on a small scale, maybe in a unit of the hospital before it is implemented on a large scale. This provides an opportunity to identify loopholes in the implementation plan and make necessary changes, potentially increasing the program's likelihood of success on a larger scale.

When stewardship interventions possess these attributes, they are easily embraced and are less likely to face rigid opposition when implemented.

3.3.5 Identify change agents/stewardship champions

To effectively initiate long-lasting change, local change agents are required. These change agents (antimicrobial stewardship champions) are early adopters of the changes initiated by the AMS team. These champions lead others to adopt new practices, and transmit their commitment and enthusiasm. These champions are often well respected and have the credibility and capacity to influence others. A change agent could be a senior manager or consultant clinician working to drive the implementation of, and adherence to stewardship principles among physicians in a health facility. Change agents should be able to effectively navigate the cultural, individual, and social differences among staff that may limit the success of the ASP. They should be equipped with knowledge of antimicrobial stewardship practices and leadership skills to sustain change.

The activities of these change agents should receive positive reinforcement and support from the antimicrobial stewardship team and health facility leadership, they should be acknowledged and appreciated. These agents can also be channels of feedback on the progress of the ASP.

3.3.6 Provide a favourable environment for change

All health workers must be educated on antimicrobial stewardship practices and their benefits to them and the patients. Education can be done using clinical case studies, lectures, and e-learning materials. This can be integrated into staff mandatory development programs. When possible, a trip to another health facility where ASPs are in full effect is encouraged. Through site visits and study tours, they see for themselves changes to expect through an ASP and the real-life benefits to the patients and health practitioners. The stewardship team and health leadership should clearly state the roles and responsibilities of the different departments and staff involved in the ASP. Leadership should provide the resources and support necessary to effectively execute these tasks.

ASP requires multidisciplinary collaborations, thus, structural and organisational changes and strategies should be established to promote quick communication; encourage feedback and open discussion on challenges or important issues that arise during the programs; and advance partnership across departments.

Change agents and the AS team should work to provide continuous encouragement and technical assurance to staff that are involved in the day-to-day activities of the AS processes. Health leadership must positively enforce their commitment to change processes initiated by the AS team. Their attitudes and statement should clearly communicate "This change is necessary, and we support it".

It's commonly said that "perfect is the enemy of good", it is important that ASPs are initiated as soon as possible and are not delayed till when the situation is "perfect".

3.3.7 Monitor and evaluate stewardship programs

Adherence to stewardship interventions should be monitored and evaluated periodically. ASP can be evaluated by the stewardship team every 3–6 months (or at a frequency agreed upon by the stewardship team), and the impact can be shared with stakeholders to further obtain their support and approval of the program. Results should also be discussed with the health workforce and various departments, to provide possible solutions to challenges encountered while initiating the stewardship interventions.

Antimicrobial prescription and antimicrobial use audits should be initiated regularly, antibiograms and drug formularies should be updated regularly based on data from antimicrobial resistance and susceptibility surveillance. Monitoring and evaluation practices provide an opportunity to obtain feedback from clinicians and stakeholders on ASP, address challenges and encourage further compliance with stewardship interventions. SMART goals and indicators agreed on at the beginning of the intervention process will be valuable during monitoring and evaluation. Information regarding the success of ASP, and patterns of antimicrobial susceptibility can be shared with regional and national surveillance teams and health authorities.

3.3.8 Implement feedback

Important information gotten from the monitoring and evaluation process, and antibiotic audits should be shared with the health professionals and leadership and should be implemented. Identified challenges should be discussed and practical solutions should be initiated.

4. Conclusion

The problem of antimicrobial resistance possesses a global threat to health, finance, and food security, and although only a few "new" antimicrobials have been developed over the years, a combination of effective strategies have been identified to mitigate this global public health challenge. Antimicrobial stewardship programs targeted at implementing evidence-based antimicrobial use, restricting the use of "broad-spectrum" and "reserve" antimicrobials, limiting the occurrence of AMR-related infections, and promoting infection prevention and control practices should be implemented in all settings where antimicrobials are prescribed and used. A "One health" approach to tackle AMR should be prioritised considering the interaction and interdependence between and among humans, animals, and the environment. Stewardship interventions have been proven to be effective in improving antimicrobial use and patient outcomes, and are readily accepted and implemented when they are presented to be beneficial to the health practitioner and the patient; made easy to understand and implement; and unequivocally supported by health leadership and important stakeholders.

5. Additional resources

5.1 ASP in the hospital

- Vieceli T, Rello J. Optimisation of antimicrobial prescription in the hospital. *European Journal of Internal Medicine*. 2022. https://doi.org/10.1016/j.ejim.2022.08.035.

- Mendelson M, Morris AM, Thursky K, Pulcini C. How to start an antimicrobial stewardship program in a hospital. *Clinical Microbiology and Infection*. 2020;26(4):447–453. https://doi.org/10.1016/j.cmi.2019.08.007.

- Cheong HS, Park KH, Kim HB, Kim SW, Kim B, Moon C, Lee MS, Yoon YK, Jeong SJ, Kim YC, Eun BW. Core elements for implementing antimicrobial stewardship programs in Korean General Hospitals. *Infectious Chemotherapy*. 2022;54(4). doi: 10.3947/ic.2022.0171.

- Doron S, Davidson LE. Antimicrobial stewardship. *Mayo Clinic Proceedings*. 2011;86(11):1113–1123. https://doi.org/10.4065/mcp.2011.0358.

5.2 ASP in the ICU

- Chiotos K, Tamma PD, Gerber JS. Antibiotic stewardship in the intensive care unit: Challenges and opportunities. *Infectious Control Hospital Epidemiology*. 2019;40(6):693–698. doi: 10.1017/ice.2019.74.

- Murphy CV, Reed EE, Herman DD, Magrum B, Beatty JJ, Stevenson KB. Antimicrobial Stewardship in the ICU. *Seminars in Respiratory Critical Care Medicine*. 2022;43(1):131–140. doi: 10.1055/s-0041-1,740,977.

- Pickens CI, Wunderink RG. Principles and Practice of Antibiotic Stewardship in the ICU. *Chest*. 2019;156(1):163–171. doi: 10.1016/j.chest.2019.01.013.

5.3 ASP in nursing homes

- Crnich CJ, Jump R, Trautner B, Sloane PD, Mody L. Optimising antibiotic stewardship in nursing homes: A narrative review and recommendations for improvement. *Drugs Ageing*. 2015;32(9):699–716. doi: 10.1007/s40266-015-0292-7.

- McElligott M, Welham G, Pop-Vicas A, Taylor L, Crnich CJ. Antibiotic stewardship in nursing facilities. *Infectious Disease in Clinical North America*. 2017;31(4):619–638. doi: 10.1016/j.idc.2017.07.008.

Author details

Olanrewaju O. Popoola[1,2]

1 Obafemi Awolowo University, Ile-Ife, Nigeria

2 HealthPlus Pharmacy Limited, Nigeria

*Address all correspondence to: popoolalanre8@gmail.com

References

[1] British Society for Antimicrobial Chemotherapy. Antimicrobial stewardship from principles to practice. 2018. Available from: https://www.bsac.org.uk/antimicrobialstewardshipebook/BSAC-AntimicrobialStewardship-FromPrinciplestoPractice-eBook.pdf [Accessed: April 25, 2023]

[2] Murray CJ, Ikuta KS, Sharara F, Swetschinski L, Robles AG, Gray A, et al. Global burden of bacterial antimicrobial resistance in 2019: A systematic analysis. The Lancet. 2022;**399**(10325):629-655. DOI: 10.1016/S0140-6736(21)02724-0

[3] World Bank Group. Drug-Resistant Infections: A Threat to Our Economic Future. Washington, DC: World Bank Group; 2017. Available from: https://www.worldbank.org/en/topic/health/publication/drug-resistant-infections-a-threat-to-our-economic-future [Accessed: May 3, 2023]

[4] World Health Organization (WHO). Antimicrobial Stewardship Interventions: A Practical Guide. Copenhagen: WHO Regional Office for Europe; 2021. Available from: https://apps.who.int/iris/handle/10665/340709 [Accessed: May 3, 2023]

[5] Ayukekbong JA, Ntemgwa M, Atabe AN. The threat of antimicrobial resistance in developing countries: Causes and control strategies. Antimicrobial Resistance and Infection Control. 2017;**6**:47. DOI: 10.1186/s13756-017-0208-x

[6] WHO. Global Action Plan on Antimicrobial Resistance. Geneva: World Health Organization; 2015. Available from: https://www.who.int/publications/i/item/9789241509763 [Accessed: May 7, 2023]

[7] Bankar NJ, Ugemuge S, Ambad RS, Hawale DV, Timilsina DR. Implementation of antimicrobial stewardship in the healthcare setting. Cureus. 2022;**14**(7):e26664. DOI: 10.7759/cureus.26664

[8] Catho G, Centemero NS, Catho H, Ranzani A, Balmelli C, Landelle C, et al. Factors determining the adherence to antimicrobial guidelines and the adoption of computerised decision support systems by physicians: A qualitative study in three European hospitals. International Journal of Medical Informatics. 2020;**141**:104233. DOI: 10.1016/j.ijmedinf.2020.104233

[9] Morency-Potvin P, Schwartz DN, Weinstein RA. Antimicrobial stewardship: How the microbiology laboratory can right the ship. Clinical Microbiology Reviews. 2017;**30**:381-407. DOI: 10.1128/CMR.00066-16

[10] Bharadwaj A, Rastogi A, Pandey S, Gupta S, Sohal JS. Multidrug-resistant Bacteria: Their mechanism of action and prophylaxis. BioMed Research International. 2022;**2022**:5419874. DOI: 10.1155/2022/5419874

[11] Tonazzi S, Prenovost L, Scheuermann S. Delayed antibiotic prescribing to reduce antibiotic use: An urgent care practice change. BMJ Open Quality. 2022;**11**:e001513. DOI: 10.1136/bmjoq-2021-001513

[12] Llor C, Moragas A, Cots JM. Implementation of the delayed antibiotic prescribing strategy. Prospective observation study in primary care. Revista Espanola de Quimioterapia. 2022;**35**(2):213-217 10.37201/req/141.2021

[13] Ghebrehewet S, Shepherd W, Panford-Quainoo E, Shantikumar S, Decraene V, Rajendran R, et al. Implementation of a delayed prescribing model to reduce antibiotic prescribing for suspected upper respiratory tract infections in a hospital outpatient department, Ghana. Antibiotics (Basel). 2020;**9**(11):773. DOI: 10.3390/antibiotics9110773

[14] National Centre for Antimicrobial Stewardship (NCAS). Antimicrobial Stewardship. 2020. Available from: https://www.ncas-australia.org/antimicrobial-formulary-and-restrictions [Accessed: May 10, 2023]

[15] Public Health Ontario (PHO). Antimicrobial Stewardship Strategy: Formulary restriction. Available from: https://www.publichealthontario.ca/apps/aspstrategies/data/pdf/ASP_Strategy_Formulary_Restriction.pdf [Accessed: May 15, 2023]

[16] WHO. Executive Summary: The Selection and Use of Essential Medicines 2019. Geneva: World Health Organization; 2019. Available from: https://apps.who.int/iris/handle/10665/325773

[17] WHO. The WHO AWaRe (Access, Watch, Reserve) Antibiotic Book. Geneva: World Health Organization; 2022. Available from: https://www.who.int/publications/i/item/WHO-MHP-HPS-EML-2022.02

[18] Taplitz RA, Ritter ML, Torriani FJ. Infection prevention and control, and antimicrobial stewardship. Infectious Diseases. 2017;**2017**:54-61. DOI: 10.1016/B978-0-7020-6285-8.00006-X

[19] Kpokiri EE, Taylor DG, Smith FJ. Development of antimicrobial stewardship programs in low and middle-income countries: A mixed-methods study in Nigerian Hospitals. Antibiotics (Basel). 2020;**9**(4):204. DOI: 10.3390/antibiotics9040204

[20] Pulcini C, Beovic B, Béraud G, Carlet J, Cars O, Howard P, et al. Ensuring universal access to old antibiotics: A critical but neglected priority. Clinical Microbiology and Infection. 2017;**23**(9):590-592. DOI: 10.1016/j.cmi.2017.04.026

[21] Pan American Health Organization (PAHO). Florida International University. Recommendations for Implementing Antimicrobial Stewardship Programs in Latin America and the Caribbean: Manual for Public Health Decision-Makers. Washington, D.C.: PAHO; 2018. Available from: https://iris.paho.org/handle/10665.2/49645

[22] Malan L, Labuschagne Q, Brechtelsbauer E, Goff DA, Schellack N. Sustainable access to antimicrobials; a missing component to antimicrobial stewardship—A tale of two countries. Frontiers in Public Health. 2018;**6**:324. DOI: 10.3389/fpubh.2018.00324

[23] Dyar OJ, Huttner B, Schouten J, Pulcini C, ESGAP (ESCMID Study Group for Antimicrobial stewardshiP). What is antimicrobial stewardship? Clinical Microbiology and Infection. 2017;**23**(11):793-798. DOI: 10.1016/j.cmi.2017.08.026

[24] McEwen SA, Collignon PJ. Antimicrobial resistance: A one health perspective. Microbiology Spectroscopy. 2018;**6**(2):1-13. DOI: 10.1128/microbiolspec.ARBA-0009-2017

[25] Chowdhury S, Rheman S, Debnath N, Delamare-Deboutteville J, Akhtar Z, Ghosh S, et al. Antibiotics usage practices in aquaculture in Bangladesh and their associated

factors. One Health. 2022;**15**:100445. DOI: 10.1016/j.onehlt.2022.100445

[26] Larsson DGJ, Flach CF. Antibiotic resistance in the environment. Nature Reviews. Microbiology. 2022;**20**(5):257-269. DOI: 10.1038/s41579-021-00649-x

[27] Sagaseta de Ilurdoz M, Sadhwani JJ, Reboso JV. Antibiotic removal processes from water & wastewater for the protection of the aquatic environment - a review. JWPE. 2022;**2022**:45. DOI: 10.1016/j.jwpe.2021.102474

[28] Goulas A, Livoreil B, Grall N, et al. What are the effective solutions to control the dissemination of antibiotic resistance in the environment? A systematic review protocol. Environmental Evidence. 2018;7:3. DOI: 10.1186/s13750-018-0118-2

[29] CDC. Core Elements of Hospital Antibiotic Stewardship Programs. Atlanta, GA: US Department of Health and Human Services, CDC; 2019. Available from: https://www.cdc.gov/antibiotic-use/core-elements/hospital.html. Accessed May 10th 2023.

[30] Zembles TN, Nakra N, Parker SK. Extending the reach of antimicrobial stewardship to pediatric patients. Infectious Disease and Therapy. 2022;**11**(1):101-110. DOI: 10.1007/s40121-022-00590-3

[31] McCreary ML, Tse-Chang A, Forbes KL, Foulds JL. Physician experiences implementing antimicrobial stewardship rounds in pediatric hospital medicine: An exploratory, qualitative study. Antimicrobial Steward Healthcare Epidemiology. 2021;**1**(1):e11. DOI: 10.1017/ash.2021.175

[32] Renk H, Sarmisak E, Spott C, Kumpf M, Hofbeck M, Hölzl F. Antibiotic stewardship in the PICU: Impact of ward rounds led by paediatric infectious diseases specialists on antibiotic consumption. Scientific Reports. 2020;**10**(1):8826. DOI: 10.1038/s41598-020-65671-0

[33] Hwang S, Kwon KT. Core elements for successful implementation of antimicrobial stewardship programs. Infectious Chemotherapy. 2021;**53**(3):421-435. DOI: 10.3947/ic.2021.0093

[34] Royal Pharmaceutical Society. The Pharmacy Contribution to Antimicrobial Stewardship. 2017. Available from: https://www.rpharms.com/Portals/0/RPS%20document%20library/Open%20access/Policy/AMS%20policy.pdf [Accessed: May 17, 2023].

[35] Jantarathaneewat K, Camins B, Apisarnthanarak A. The role of the clinical pharmacist in antimicrobial stewardship in Asia: A review. Antimicrobial Steward Healthcare Epidemiology. 2022;**2**(1):e176. DOI: 10.1017/ash.2022.310

[36] Mishima Y, Nawa N, Asada M, Nagashima M, Aiso Y, Nukui Y, et al. Impact of antibiotic time-outs in multidisciplinary ICU rounds for antimicrobial stewardship program on patient survival: A controlled before-and-after study. Crital Care Exploration. 2023;**5**(1):e0837. DOI: 10.1097/cce.0000000000000837

[37] Wirtz AL, Burns AN, Lee BR, Frank TS, Fitzmaurice L, Ogden RK, et al. Effectiveness and safety of mandatory antimicrobial indications and durations and a pharmacist-driven 48-hour time-out in a pediatric hospital. American Journal of Health-System Pharmacy. 2020;**77**(8):614-621. DOI: 10.1093/ajhp/zxaa029

[38] Paulson CM, Handley JF, Dilworth TJ, et al. Impact of a systematic pharmacist-initiated antibiotic time-out intervention for hospitalized adults. Journal of Pharmacy Practice. 2022;**35**(3):388-395. DOI: 10.1177/0897190020980616

[39] Satterfield J, Miesner AR, Percival KM. The role of education in antimicrobial stewardship. The Journal of Hospital Infection. 2020;**105**(2):130-141

[40] Sahra S, Jahangir A, De Chavez V. Antimicrobial stewardship: A review for internal medicine physicians. Cureus. 2021;**13**(4):e14385. DOI: 10.7759/cureus.14385

[41] Ostrowsky B, Banerjee R, Bonomo RA, Cosgrove SE, Davidson L, Doron S, et al. Infectious diseases physicians: Leading the way in antimicrobial stewardship. Clinical Infectious Diseases. 2018;**66**(7):995-1003. DOI: 10.1093/cid/cix1093

[42] Van Huizen P, Kuhn L, Russo PL, Connell CJ. The nurses' role in antimicrobial stewardship: A scoping review. International Journal of Nursing Studies. 2021;**113**:103772. DOI: 10.1016/j.ijnurstu.2020.103772

[43] Gotterson F, Buising K, Manias E. Nurse role and contribution to antimicrobial stewardship: An integrative review. International Journal of Nursing Studies. 2021;**117**:103787. DOI: 10.1016/j.ijnurstu.2020.103787

[44] Palavecino EL, Williamson JC, Ohl CA. Collaborative antimicrobial stewardship: Working with microbiology. Infectious Disease Clinics of North America. 2020;**34**(1):51-65. DOI: 10.1016/j.idc.2019.10.006

[45] WHO. Antimicrobial Stewardship Programs in Health-Care Facilities in Low- and Middle-Income Countries. A Practical Toolkit. Geneva: World Health Organization; 2019. Available from: https://apps.who.int/iris/handle/10665/329404

Chapter 5

Antimicrobial Stewardship in Resource-Limited Settings

Simeon Chijioke Amadi, Faithful Miebaka Daniel, Sokeipirim Ikiroma and Ibinabo Laura Oboro

Abstract

Antimicrobials are very important for the treatment of patients. With decades of haphazard prescription and inappropriate use, antimicrobial resistance (AMR) has emerged as a global public health threat. In resource-limited settings, besides AMR, there is also the issue of access to antibiotics and increased healthcare costs. In the past, the discovery of new drugs and the modification of older ones helped to counter antimicrobial resistance. Since the 1970s, only a handful of new agents (a few of which are novel) have been introduced into clinical practice periodically. This makes the existing antibiotics to be a limited resource. Antibiotics must, therefore, be used more responsibly and managed carefully to extend their usefulness while being made available to the patients who truly need them. Antimicrobial stewardship (AMS) refers to systematic actions or interventions that inculcate sustainable, rational, and evidence-based antibiotic prescription and use patterns in healthcare institutions. Implementation of AMS programmes would be an appropriate step towards improving patient outcomes, reducing AMR and its adverse economic impact, and building the best-practices capacity of healthcare professionals, among others. Despite these seemingly beneficial benefits, the implementation of antimicrobial stewardship programmes (ASPs) remains relatively limited and faces complex challenges in resource-poor healthcare settings.

Keywords: antimicrobial, stewardship, resource, limited, setting

1. Introduction

Today's world is a product of continuous change driven by research, industry and innovation to address humanity's greatest needs. The development of antimicrobials has been an ongoing process dating back to ancient civilisations that utilised plant extracts and mouldy bread to treat infections [1]. Alexander Fleming's serendipitous discovery of penicillin in 1928 ushered in an era of growing interest in antimicrobial formulation and its ability to save millions of lives [1]. Since then, synthetic chemistry and modern drug discoveries have become the norm. However, the world faces an even more significant threat that may upend the progress made over the years: antimicrobial resistance [1, 2].

Scientists have striven to utilise research and innovation to combat emerging drug resistance and the evolving challenges of infectious diseases [2]. Antimicrobial resistance (AMR) is a primary global public health concern. It occurs when microorganisms develop the ability to evade the effects of antimicrobial medication, resulting in a reduction or loss of the medications' effectiveness in treating infections [2]. The main drivers of AMR are the inappropriate use and overuse of antimicrobials, which can stem from factors such as poor infection control practices, inadequate surveillance systems, improper prescription practices, inadequate patient education, limited diagnostic facilities, unauthorised sale of antimicrobials, lack of appropriate functioning drug regulatory mechanisms, and non-human use of antimicrobials such as in animal production [2, 3]. The challenges posed by AMR require a comprehensive and coordinated approach to promote responsible antimicrobial use, develop new drugs, strengthen infection prevention and control practices, and enhance surveillance [3]. To this end, antimicrobial stewardship is a crucial strategy that aims to optimise the use of antimicrobial medicines while minimising the development of AMR. It involves a multidisciplinary approach that includes healthcare professionals, policymakers, and patients [4].

Addressing the global threat of resistant infections requires a concerted effort to tackle AMR and promote antimicrobial stewardship. Antimicrobial stewardship is the practice of optimising the use of antimicrobial drugs to combat the growing issue of antimicrobial resistance [4]. It is essential to ensure the responsible use of these drugs. However, resource-poor settings face unique challenges in implementing effective antimicrobial stewardship programmes due to limited resources, infrastructure, and funding [3]. These challenges include limited access to healthcare, inadequate diagnostic capabilities, scarcity of essential medicines, and difficulties in implementing infection prevention and control measures [5]. Innovative solutions and a commitment to ensuring access to essential medicines and healthcare services for all are necessary to address these challenges.

2. The significance of antimicrobial stewardship

Addressing antimicrobial stewardship in resource-poor settings is vital as it directly affects vulnerable populations, decreases the economic burden of inappropriate antimicrobial use, and controls the spread of antimicrobial resistance [5–7]. In this chapter, we will delve deeper into the topic, exploring the specific challenges faced by resource-poor settings and underscoring the importance of implementing effective antimicrobial stewardship programmes.

Antimicrobial stewardship programmes in resource-poor settings face several unique challenges that hamper their effectiveness. One of the critical challenges is the limited access to healthcare. In many low- and middle-income countries, most people live in rural areas with scarce healthcare facilities. This limited access to healthcare facilities makes it difficult for patients to receive proper diagnosis and treatment for infectious diseases, leading to the inappropriate use of antimicrobials and the development of resistance [5–7].

Another significant challenge is inadequate diagnostic capabilities. Diagnostic tests such as blood cultures, microbiological cultures, and susceptibility testing are essential for identifying the causative agent of an infection and determining the appropriate antimicrobial therapy [5–7]. However, many resource-poor settings lack the necessary diagnostic capabilities to perform these tests, leading to the widespread use of broad-spectrum antimicrobials and the development of resistance [5–7].

The scarcity of essential medicines is another challenge faced by resource-poor settings. Essential drugs, including antimicrobials, are often unavailable or in short supply in these settings due to a lack of funding, poor infrastructure, and limited supply chains [5–7]. This scarcity of essential medicines can lead to inappropriate antimicrobials, contributing to the development of resistance [5–7]. Implementing infection prevention and control measures significantly challenges antimicrobial stewardship in resource-poor settings [5–7]. Lack of basic amenities such as clean water, sanitation facilities, and waste management systems can rapidly spread infectious diseases, making it difficult to control outbreaks.

To overcome these challenges, effective antimicrobial stewardship programmes must be implemented in resource-poor settings. These programmes should focus on improving access to healthcare, strengthening diagnostic capabilities, ensuring the availability of essential medicines, and implementing infection prevention and control measures [8]. Improving access to healthcare can be accomplished by deploying community health workers who can provide critical diagnostic and treatment services to patients in rural areas [5]. Rapid diagnostic tests that do not require sophisticated laboratory facilities can also help improve diagnostic capabilities in resource-poor settings [7]. Ensuring the availability of essential medicines requires a coordinated effort between governments, international organisations, and the private sector [6]. This effort should focus on improving supply chain management, increasing funding for research and development of new antimicrobial drugs, and reducing the cost of existing medications [7]. Implementing infection prevention and control measures requires basic amenities such as clean water, sanitation facilities, and waste management systems. These amenities can be provided through public–private partnerships focusing on improving infrastructure in resource-poor settings [5].

To sum up, antimicrobial stewardship is crucial in addressing the global threat of antimicrobial resistance. Nonetheless, implementing effective antimicrobial stewardship programmes in resource-poor settings requires a comprehensive and coordinated effort to address these settings' unique challenges. Improving access to healthcare, strengthening diagnostic capabilities, ensuring the availability of essential medicines, and implementing infection prevention and control measures are vital steps towards achieving this goal.

3. Understanding antimicrobial resistance in resource-poor settings

The issue of antimicrobial resistance (AMR) is a pressing global public health crisis with far-reaching implications for healthcare systems worldwide [9]. The emergence and spread of antimicrobial resistance significantly threaten our ability to treat infectious diseases effectively, increasing morbidity, mortality, and healthcare costs [10, 11]. The challenges in combating AMR are even more pronounced in resource-poor settings, characterised by poor resources, limited infrastructure, and scarce funding [10, 12, 13]. These settings face unique obstacles in addressing AMR, including limited access to healthcare, inadequate diagnostic capabilities, scarcity of essential medicines, and challenges in implementing infection prevention and control measures [10, 12–14]. These factors contribute to inappropriate use of antimicrobials, such as overprescribing, inadequate dosing, and misuse, leading to severe consequences, including treatment failures, increased morbidity and mortality, and disproportionate impacts on vulnerable populations [10, 12–14]. Addressing AMR in resource-poor settings is, therefore, mitigating the effects and preventing the further spread of resistance.

However, the challenges in addressing antimicrobial resistance in resource-poor settings are further compounded by the fragility of health systems. Fragile health systems, characterised by weak characteristics, limited resources, and inadequate governance, pose a significant barrier to effectively combating antimicrobial resistance [15]. The fragility of these health systems undermines efforts to address AMR and poses challenges in addressing other health priorities and achieving global health goals. Many resource-poor settings struggle with fragile health systems, further hampering their ability to address AMR [15, 16]. The World Health Organisation (WHO) has identified six building blocks of health systems: leadership and governance, health workforce, health information systems, access to essential medicines, financing, and service delivery [17]. Addressing these building blocks is critical for improving healthcare infrastructure, human resources, and laboratory capacity, which is necessary for managing antimicrobial resistance in these settings [17].

Therefore, comprehensive strategies should be implemented to improve access to healthcare, enhance diagnostic capabilities, ensure the availability of essential medicines, and strengthen infection prevention and control measures. Additionally, capacity building and education initiatives are vital to promote responsible antimicrobial use and raise awareness about the significance of AMR [9]. By addressing both the challenges in resource-poor settings and the fragility of health systems, we can contribute to safeguarding the effectiveness of antimicrobials, protecting vulnerable populations, and preserving the efficacy of these life-saving drugs for future generations.

4. Challenges and barriers to antimicrobial stewardship in resource-poor settings

The global healthcare context in resource-poor settings is often characterised by limited resources, infrastructure, and funding to adequately meet the population's healthcare needs [15, 17]. This scenario poses unique challenges when it comes to addressing antimicrobial stewardship. Limited access to healthcare facilities and professionals, inadequate diagnostic capabilities, scarcity of essential medicines, and difficulties implementing infection prevention and control measures are some of the challenges faced in these settings [18].

The lack of reliable and timely diagnostic tools hampers accurate determination of the cause of infection and appropriate antimicrobial therapy selection [19]. Addressing antimicrobial stewardship in resource-poor settings is essential due to its impact on vulnerable populations, the economic burden of inappropriate antimicrobial use, and the potential for spreading antimicrobial resistance [16, 19]. Inadequate stewardship can result in treatment failures, increased morbidity and mortality, and disproportionately affect vulnerable populations. The emergence and spread of antimicrobial resistance are particularly concerning in resource-poor settings, where limited access to healthcare, inadequate diagnostics, and infection prevention and control challenges can contribute to its development [5, 19].

The limitations to antimicrobial stewardship in resource-poor settings can be attributed to several factors, including the lack of trained healthcare professionals and awareness, financial constraints, limited resources, and cultural and behavioural factors affecting antimicrobial use [16, 19]. The need for more trained healthcare professionals in resource-poor settings poses a significant challenge to implementing effective antimicrobial stewardship programmes. Financial constraints and limited resources also play a crucial role in inhibiting antimicrobial stewardship efforts.

Cultural beliefs, practices, and patient expectations may influence the demand for antimicrobials, leading to their inappropriate use [20]. For instance, in some cultures, antibiotics are considered a miracle cure, and their use is expected for any illness, regardless of whether bacteria or viruses cause it [20]. Additionally, limited health literacy and understanding of the risks of antimicrobial misuse among community members can perpetuate the problem. Therefore, addressing these factors is crucial in implementing effective antimicrobial stewardship programmes in resource-poor settings.

Several interventions can be implemented to ensure effective antimicrobial stewardship in resource-poor settings. Firstly, there is a need for increased investment in healthcare infrastructure, particularly in diagnostic capabilities and infection prevention and control measures [19]. This investment should be accompanied by appropriate training of healthcare workers on antimicrobial stewardship and infection control practices. Secondly, there is a need for increased awareness campaigns to educate the public on the proper use of antimicrobials and the risks associated with their misuse [14].

Thirdly, there is a need to develop guidelines and protocols for the appropriate use of antimicrobials in resource-poor settings. Such policies should be context-specific, considering the unique challenges faced in resource-poor settings [5, 19]. Fourthly, there is a need for increased surveillance of antimicrobial use and resistance patterns to guide the development and implementation of effective stewardship programmes [18, 21].

In conclusion, antimicrobial stewardship in resource-poor settings is crucial to prevent the emergence and spread of antimicrobial resistance and ensure the appropriate use of antimicrobials. Addressing the challenges these settings face requires a multi-pronged approach that involves increased investment in healthcare infrastructure, proper training of healthcare workers, increased public awareness campaigns, and developing context-specific guidelines and protocols. These interventions will go a long way in ensuring effective antimicrobial stewardship in resource-poor settings.

5. Strategies for implementing antimicrobial stewardship in resource-poor settings

Antimicrobial stewardship programmes (ASPs) are crucial in promoting responsible antimicrobial use and combating AMR. However, implementing effective ASPs in resource-poor settings is challenging due to the need for more resources and infrastructure. To address these challenges, multiple aspects of healthcare delivery must be addressed. Building local capacity and training healthcare professionals is crucial [22–24]. This means providing education and training on antimicrobial stewardship principles, appropriate use, and infection prevention and control measures [25]. It is also essential to strengthen healthcare infrastructure, including improving access to diagnostics such as laboratory facilities and point-of-care tests [5, 26, 27]. This enables accurate diagnosis of infections and facilitates appropriate antimicrobial prescribing. Developing guidelines and protocols for antimicrobial use tailored to the specific context of resource-poor settings helps standardise practices and promote responsible prescribing [4, 25, 26].

ASPs also require a cultural shift in the healthcare system. Promoting rational prescribing practices and patient education is crucial in shifting cultural and behavioural factors contributing to antimicrobial misuse [20]. Public awareness campaigns and educating patients about the risks of inappropriate antimicrobial use and the importance of completing prescribed treatments are vital strategies [23].

Engaging stakeholders, including healthcare professionals, policymakers, and community leaders, fosters collaboration and creates a supportive environment for implementing ASPs [28–30]. Experts in the field have identified nine key strategies to enhance antimicrobial stewardship (AMS) [28]. These include implementing quality improvement measures, promoting peer learning across different disciplines, assigning AMS leads, carrying out individual-level prescribing audits, creating practical tools for prescribing audits, improving induction processes for new prescribers, standardising local approaches to antibiotic prescribing, providing online AMS training to all patient-facing staff, and increasing staff time devoted to AMS work through the standardisation of AMS-related roles [28].

Implementing ASPs in resource-poor settings is challenging and requires a comprehensive and coordinated approach. Developing national action plans and policies can provide a framework for implementing ASPs and ensuring sustainability [31, 32]. Additionally, funding and technical support from international organisations and donor agencies can provide critical resources to support ASP implementation [25, 33, 34].

In conclusion, addressing AMR in resource-poor settings requires a comprehensive approach that includes strengthening healthcare infrastructure, improving access to essential medicines, enhancing diagnostic capabilities, and implementing effective infection prevention and control measures. Capacity building and education for healthcare professionals and the community are vital in promoting responsible antimicrobial use. Promoting rational prescribing practices and patient education is crucial in shifting cultural and behavioural factors contributing to antimicrobial misuse. Engaging stakeholders and developing national action plans and policies can provide a framework for implementing ASPs and ensuring sustainability [31, 32]. Like the National Action Plan for AMR developed in India [35]. In 2019, a framework for addressing AMR was designed to be an ongoing and adaptable process that responds to contextual factors, allowing for continuous improvement and refinement of national action plans [36]. This was to address the need for a systematic approach to the governance of national plans [36]. By systematically addressing these strategies, resource-poor settings can enhance their ability to combat antimicrobial resistance and improve patient outcomes.

6. Lessons from case studies

The successful implementation of antimicrobial stewardship programmes in resource-poor settings has been the subject of various studies. These studies have demonstrated the potential for positive outcomes, such as significantly reducing inappropriate antimicrobial prescribing and improving guideline adherence [26, 35]. The key to success lies in tailored training and education programmes considering resource-poor settings' specific contexts and challenges [26]. Additionally, developing and implementing evidence-based local guidelines and protocols can considerably reduce the prevalence of antimicrobial resistance and improve patient outcomes. Continuous monitoring and feedback to healthcare professionals also play a vital role in ensuring responsible antimicrobial use [18, 21]. These successful initiatives challenge the notion that limited resources are insurmountable barriers to implementing effective antimicrobial stewardship programmes in resource-poor settings.

Numerous studies have explored the implementation of antimicrobial stewardship programmes in resource-poor settings, revealing their potential to reduce inappropriate antimicrobial prescribing and improve adherence to guidelines. The key to success lies in tailored training and education programmes that address the unique challenges of such settings [15, 22, 23, 31]. Evidence-based local procedures and protocols are crucial in tackling antimicrobial resistance and enhancing patient outcomes.

Despite limited resources, establishing active antimicrobial stewardship programmes in resource-poor settings is achievable through a collaborative approach involving healthcare professionals, policymakers, and the community [30]. It is essential to provide tailored training and education programmes that cater to the specific challenges faced by healthcare professionals [30]. Developing and implementing context-specific guidelines and protocols based on evidence can considerably reduce antimicrobial resistance and improve patient outcomes [20]. Regular monitoring and feedback to healthcare professionals are critical for responsible antimicrobial use.

The successful implementation of antimicrobial stewardship programmes in resource-poor settings challenges the notion that limited resources are insurmountable barriers. By adopting an evidence-based, context-specific approach, healthcare professionals and policymakers can effectively address the issue of antimicrobial resistance in such settings.

7. Overcoming challenges and sustainability of antimicrobial stewardship programmes

Antimicrobial resistance (AMR) is a pressing global health issue that threatens the effective prevention and treatment of infections caused by bacteria, viruses, fungi, and parasites [9]. The misuse and overuse of antimicrobial agents, poor infection prevention and control practices, and inadequate access to clean water, sanitation, and hygiene are major drivers of AMR. Resource-poor settings, where access to healthcare and essential medicines is limited, are particularly vulnerable to the negative impact of AMR. Antimicrobial stewardship programmes (ASPs) are crucial interventions to address AMR and promote the rational use of antimicrobial agents [31]. However, the implementation and sustainability of ASPs in resource-poor settings face several challenges that must be addressed [31]. Financial constraints and limited resources are significant obstacles to implementing and sustaining ASPs in resource-poor settings. External funding from international organisations, donor organisations, and philanthropic foundations can help overcome these constraints [33, 34]. Advocating for increased healthcare budgets and prioritising allocating ASPs can also contribute to addressing and identifying innovative approaches for funding, resource mobilisation and partnerships, which can provide additional resources and expertise [37].

Partnerships with organisations, institutions, and pharmaceutical companies can also provide access to expertise, resources, and funding. Collaborating with local communities and engaging community leaders can promote sustainability by fostering ownership and support for ASPs [30]. Engaging patients and their families in antimicrobial stewardship efforts can also increase awareness and guideline adherence [23]. Integrating ASPs into existing healthcare systems is crucial for long-term sustainability [17, 24, 38]. This can be achieved by incorporating antimicrobial stewardship principles and training into medical and nursing curricula and establishing dedicated antimicrobial stewardship teams or committees within healthcare facilities [24]. Collaboration between healthcare professionals, policymakers, and regulators is essential to develop and enforce antimicrobial prescribing guidelines and policies that align with national and international recommendations [38]. Furthermore, leveraging existing healthcare infrastructure and systems, such as electronic medical records and surveillance systems, can facilitate monitoring and evaluating antimicrobial use and resistance patterns [21].

Ensuring the long-term sustainability and scalability of ASPs is a critical consideration. This can be achieved by establishing monitoring and evaluation mechanisms to assess the programmes' impact and identify areas for improvement [38]. Continuous education and training of healthcare professionals on antimicrobial stewardship principles and practices is essential to maintain awareness and adherence to guidelines [24]. Fostering a culture of accountability and responsibility among healthcare professionals and the community is crucial for sustaining the momentum of ASPs. Engaging stakeholders, including policymakers, healthcare professionals, patients, and community members, in discussions and decision-making processes can contribute to the sustainability and scalability of ASPs [27].

Finally, implementing and sustaining ASPs in resource-poor settings requires addressing frequent constraints, forming partnerships, integrating stewardship into existing healthcare systems, and focusing on long-term scalability. When implemented in a coordinated and comprehensive manner, these strategies can contribute to the successful implementation and long-term impact of antimicrobial stewardship initiatives in resource-poor settings.

8. Future directions and recommendations

The field of antimicrobial stewardship in resource-poor settings presents several areas for further exploration. To ensure the sustained effectiveness and sustainability of antimicrobial stewardship programmes, research should investigate the long-term impact of such programmes on antimicrobial resistance patterns, patient outcomes, and healthcare costs [39]. Additionally, evaluating the effectiveness of different strategies and interventions used in antimicrobial stewardship programmes, including education and training programmes, guidelines and protocols, and electronic decision support systems, is essential [39, 40]. Understanding the social and behavioural factors influencing antimicrobial use in these settings is also crucial. Policymakers should prioritise developing and implementing national antimicrobial stewardship policies and guidelines and advocate for increased funding and resources to support programme implementation and sustainability [31, 32, 35, 36]. Global collaboration and knowledge sharing can also advance antimicrobial stewardship efforts in resource-poor settings. By addressing these areas, resource-poor locations can strengthen their antimicrobial stewardship programmes and contribute to the global fight against antimicrobial resistance.

Antimicrobial resistance (AMR) is a growing concern globally, and it is especially significant in resource-poor settings. In these locations, the prevalence of infectious diseases is high, and antibiotics are often the primary treatment option. Unfortunately, the misuse and overuse of antibiotics have led to the emergence of resistant strains of bacteria, making infections difficult to treat [6]. The World Health Organisation (WHO) has recognised the importance of antimicrobial stewardship programmes, which aim to reduce the inappropriate use of antibiotics, improve prescribing practices, and optimise patient outcomes.

One area that requires further exploration is the long-term impact of antimicrobial stewardship programmes on antimicrobial resistance patterns, patient outcomes, and healthcare costs [40, 41]. While it is clear that such programmes can reduce the development of antimicrobial resistance, the long-term effects still need to be better understood. Research should investigate the impact of stewardship programmes over extended periods to determine their sustained effectiveness and sustainability. This information can help policymakers and healthcare providers make informed decisions about resource allocation and programme implementation.

Another critical area for exploration is the evaluation of the effectiveness of different strategies and interventions used in antimicrobial stewardship programmes [42]. These may include education and training programmes, guidelines and protocols, and electronic decision support systems [42, 43]. By evaluating these strategies, healthcare providers can identify the most effective ways to reduce inappropriate antibiotic use and improve patient outcomes. This information can help providers tailor their programmes to the unique needs of their patients and communities.

Understanding the social and behavioural factors influencing antimicrobial use in resource-poor settings is also crucial [44]. While healthcare providers can significantly reduce inappropriate antibiotic use, patients and their families also have a role to play. Cultural beliefs, practices, and limited access to healthcare facilities and diagnostics can contribute to inappropriate antibiotic use [43]. By understanding these factors, healthcare providers can develop targeted interventions to address them.

In conclusion, the importance of antimicrobial stewardship in addressing the global challenge of AMR cannot be overstated. In resource-poor settings, the challenges of implementing effective stewardship programmes are significant but manageable. By addressing research, evaluation, social and behavioural factors, policymaking, and collaboration, resource-poor locations can strengthen their antimicrobial stewardship programmes and contribute to the global fight against AMR. The health of populations in resource-poor settings, as well as future generations, depends on collective action by policymakers, healthcare providers, and researchers to develop and implement effective antimicrobial stewardship programmes.

Author details

Simeon Chijioke Amadi[1*], Faithful Miebaka Daniel[2], Sokeipirim Ikiroma[1] and Ibinabo Laura Oboro[3]

1 Department of Obstetrics and Gynaecology, Rivers State University/Rivers State University Teaching Hospital, Port Harcourt, Rivers State, Nigeria

2 Community and Clinical Research Division, First On-Call Initiative, Port Harcourt, Nigeria

3 Department of Medical Microbiology, Rivers State University/Rivers State University Teaching Hospital, Port Harcourt, Rivers State, Nigeria

*Address all correspondence to: amachijio@yahoo.com

References

[1] Rahman M, Sarker SD. Antimicrobial natural products. In: Annual Reports in Medicinal Chemistry. United States: Elsevier; 2020. pp. 77-113

[2] Dyary HO, Faraj GJ, Saeed NM. History, Current Situation, and Future Perspectives on Antibiotics and Antibiotic Resistance. One Heal Triad. Vol. 2. Faisalabad, Pakistan: Unique Scientific Publisher; 2023. pp. 109-118

[3] Ayukekbong JA, Ntemgwa M, Atabe AN. The threat of antimicrobial resistance in developing countries: Causes and control strategies. Antimicrobial Resistance and Infection Control. 2017;**6**(1):1-8

[4] Fishman N. Antimicrobial stewardship. American Journal of Infection Control. 2006;**34**(5):S55-S63

[5] Kakkar AK, Shafiq N, Singh G, Ray P, Gautam V, Agarwal R, et al. Antimicrobial stewardship programs in resource-constrained environments: Understanding and addressing the need of the systems. Frontiers in Public Health. 2020;**8**:140

[6] Cox JA, Vlieghe E, Mendelson M, Wertheim H, Ndegwa L, Villegas MV, et al. Antibiotic stewardship in low-and middle-income countries: The same but different? Clinical Microbiology and Infection. 2017;**23**(11):812-818

[7] Hijazi K, Joshi C, Gould IM. Challenges and opportunities for antimicrobial stewardship in resource-rich and resource-limited countries. Expert Review of Anti-Infective Therapy. 2019;**17**(8):621-634

[8] Caron WP, Mousa SA. Prevention strategies for antimicrobial resistance: A systematic review of the literature. Infection and Drug Resistance. 2010;**3**:25-33

[9] Ferri M, Ranucci E, Romagnoli P, Giaccone V. Antimicrobial resistance: A global emerging threat to public health systems. Critical Reviews in Food Science and Nutrition. 2017;**57**(13):2857-2876

[10] Roope LSJ, Smith RD, Pouwels KB, Buchanan J, Abel L, Eibich P, et al. The challenge of antimicrobial resistance: What economics can contribute. Science (80-). 2019;**364**(6435):eaau4679

[11] Cars O, Nordberg P. Antibiotic resistance–the faceless threat. The International Journal of Risk & Safety in Medicine. 2005;**17**(3-4):103-110

[12] Tenover FC. Mechanisms of antimicrobial resistance in bacteria. The American Journal of Medicine. 2006;**119**(6):S3-S10

[13] Cosgrove SE. The relationship between antimicrobial resistance and patient outcomes: Mortality, length of hospital stay, and health care costs. Clinical Infectious Diseases. 2006;**42**(Supplement_2):S82-S89

[14] DiazGranados CA, Cardo DM, McGowan JE Jr. Antimicrobial resistance: International control strategies, with a focus on limited-resource settings. International Journal of Antimicrobial Agents. 2008;**32**(1):1-9

[15] Akilimali A, Awuah WA, Cakwira H, Mirindi MK, Masimango G, Amani T, et al. Antimicrobial resistance in Democratic Republic of Congo: The way forward. In: IJS Global Health. Vol. 6. Netherlands: Wolters Kluwer; 2023. p. e0150

[16] MacPherson EE, Reynolds J, Sanudi E, Nkaombe A, Phiri C, Mankhomwa J, et al. Understanding antimicrobial resistance through the lens of antibiotic vulnerabilities in primary health care in rural Malawi. Global Public Health. 2022;**17**(11):2630-2646

[17] World Health Organization (WHO). Monitoring the Building Blocks of Health Systems: A Handbook of Indicators and their Measurement Strategies [Internet]. Geneva, Switzerland: World Health Organization; 2010. Available from: https://www.google.com/url?sa=t&rct=j&q=&esrc=s&source=web&cd=&cad=rja&uact=8&ved=2ahUKEwij6Ifn59SAAxXgV0EAHRi8AcEQFnoECCUQAQ&url=https%3A%2F%2Fapps.who.int%2Firis%2Fbitstream%2Fhandle%2F10665%2F258734%2F9789241564052-eng.pdf&usg=AOvVaw1VUvLpCAGxolA87_WiE

[18] Iskandar K, Molinier L, Hallit S, Sartelli M, Hardcastle TC, Haque M, et al. Surveillance of antimicrobial resistance in low and middle-income countries: A scattered picture. Antimicrobial Resistance and Infection Control. 2021;**10**(1):1-19

[19] Gebretekle GB, Haile Mariam D, Abebe W, Amogne W, Tenna A, Fenta TG, et al. Opportunities and barriers to implementing antibiotic stewardship in low and middle-income countries: Lessons from a mixed-methods study in a tertiary care hospital in Ethiopia. PLoS One. 2018;**13**(12):e0208447

[20] Charani E, Smith I, Skodvin B, Perozziello A, Lucet JC, Lescure FX, et al. Investigating the cultural and contextual determinants of antimicrobial stewardship programmes across low-, middle-and high-income countries—A qualitative study. PLoS One. 2019;**14**(1):e0209847

[21] Simonsen GS, Tapsall JW, Allegranzi B, Talbot EA, Lazzari S. The antimicrobial resistance containment and surveillance approach-a public health tool. Bulletin of the World Health Organization. 2004;**82**(12):928-934

[22] Prentiss T, Weisberg K, Zervos J. Building capacity in infection prevention and antimicrobial stewardship in low- and middle-income countries: The role of partnerships inter-countries. Current Treatment Options in Infectious Diseases. 2018;**10**:7-16

[23] Virhia J, Gilmour M, Russell C, Mutua E, Nasuwa F, Mmbaga BT, et al. If you do not take the medicine and complete the dose... it could cause you more trouble: Bringing awareness, local knowledge and experience into antimicrobial stewardship in Tanzania. Antibiotics. 2023;**12**(2):243

[24] World Health Organization. Health Workers' Education and Training on Antimicrobial Resistance: Curricula Guide. Geneva, Switzerland: WHO; 2019

[25] Majumder MAA, Rahman S, Cohall D, Bharatha A, Singh K, Haque M, et al. Antimicrobial stewardship: Fighting antimicrobial resistance and protecting global public health. Infection and Drug Resistance. 2020;**13**:4713-4738

[26] Rolfe R, Kwobah C, Muro F, Ruwanpathirana A, Lyamuya F, Bodinayake C, et al. Barriers to implementing antimicrobial stewardship programs in three low-and middle-income country tertiary care settings: Findings from a multi-site qualitative study. Antimicrobial Resistance and Infection Control. 2021;**10**:1-11

[27] World Health Organization. Diagnostic Stewardship: A Guide to Implementation in Antimicrobial Resistance Surveillance Sites. Geneva,

Switzerland: World Health Organization; 2016

[28] Borek AJ, Wanat M, Sallis A, Ashiru-Oredope D, Atkins L, Beech E, et al. How can national antimicrobial stewardship interventions in primary care be improved? A stakeholder consultation. Antibiotics. 2019;**8**(4):207

[29] Wathne JS, Kleppe LKS, Harthug S, Blix HS, Nilsen RM, Charani E, et al. The effect of antibiotic stewardship interventions with stakeholder involvement in hospital settings: A multicentre, cluster randomized controlled intervention study. Antimicrobial Resistance and Infection Control. 2018;7:1-12

[30] Logan AY, Williamson JE, Reinke EK, Jarrett SW, Boger MS, Davidson LE. Establishing an antimicrobial stewardship collaborative across a large, diverse health care system. Joint Commission Journal on Quality and Patient Safety. 2019;**45**(9):591-599

[31] Frumence G, Mboera LEG, Sindato C, Katale BZ, Kimera S, Metta E, et al. The governance and implementation of the National Action Plan on antimicrobial resistance in Tanzania: A qualitative study. Antibiotics. 2021;**10**(3):273

[32] Charani E, Mendelson M, Pallett SJC, Ahmad R, Mpundu M, Mbamalu O, et al. An analysis of existing national action plans for antimicrobial resistance— Gaps and opportunities in strategies optimising antibiotic use in human populations. The Lancet Global Health. 2023;**11**(3):e466-e474

[33] Wernli D, Harbarth S, Levrat N, Pittet D. A 'whole of United Nations approach' to tackle antimicrobial resistance? A mapping of the mandate and activities of international organisations. BMJ Global Health. 2022;**7**(5):e008181

[34] Tattevin P, Levy Hara G, Toumi A, Enani M, Coombs G, Voss A, et al. Advocacy for increased international efforts for antimicrobial stewardship actions in low-and middle-income countries on behalf of alliance for the prudent use of antimicrobials (APUA), under the auspices of the International Society of Antimicrobial Chemotherapy (ISAC). Frontiers in Medicine. 2020;**7**:503

[35] Ranjalkar J, Chandy SJ. India's National Action Plan for antimicrobial resistance–an overview of the context, status, and way ahead. Journal of Family Medicine and Primary Care. 2019;**8**(6):1828

[36] Anderson M, Schulze K, Cassini A, Plachouras D, Mossialos E. A governance framework for development and assessment of national action plans on antimicrobial resistance. The Lancet Infectious Diseases. 2019;**19**(11):e371-e384

[37] Uzochukwu BSC, Ughasoro MD, Etiaba E, Okwuosa C, Envuladu E, Onwujekwe OE. Health care financing in Nigeria: Implications for achieving universal health coverage. Nigerian Journal of Clinical Practice. 2015;**18**(4):437-444

[38] World Health Organization. Antimicrobial Stewardship Interventions: A Practical Guide. Geneva, Switzerland: World Health Organization; 2021

[39] Saubolle MA. Antimicrobial resistance: Current status and future direction. American Journal of Rhinology. 2006;**20**(6):667-671

[40] Engler D, Meyer JC, Schellack N, Kurdi A, Godman B. Antimicrobial

stewardship activities in public healthcare facilities in South Africa: A baseline for future direction. Antibiotics. 2021;**10**(8):996

[41] Newland JG, Banerjee R, Gerber JS, Hersh AL, Steinke L, Weissman SJ. Antimicrobial stewardship in pediatric care: Strategies and future directions. Pharmacotherapy: The Journal of Human Pharmacology and Drug Therapy. 2012;**32**(8):735-743

[42] Same RG. The current state and future directions of inpatient pediatric antimicrobial stewardship. Infectious Disease Clinics. 2022;**36**(1):173-186

[43] Briggs DC, Oboro IL, Bob-Manuel M, Amadi SC, Enyinnaya SO, Lawson SD, et al. Antibiotic prescription patterns in paediatric wards of Rivers state university teaching hospital, southern Nigeria: A point prevalence survey. The Nigerian Health Journal. 2023;**23**(3):837-843

[44] Bassetti M, Giacobbe DR, Vena A, Brink A. Challenges and research priorities to progress the impact of antimicrobial stewardship. Drugs Context. 2019;**8**:212600

Chapter 6

Improving Antimicrobial Stewardship in Human Health Using Probiotics and Their Derivatives

Nesisa Nyathi, Duduzile Ndhlovu, Esma Rabvukwa and Abigarl Ndudzo

Abstract

Since antimicrobial resistance poses the biggest risks to human health, antimicrobial stewardship implies a strategy of responsible management. To lessen antimicrobial resistance and its impacts, a number of antimicrobial stewardship strategies are being used. One such crucial tactic is the use of probiotics and their derivatives, which directly affect the gut microbiota and have been linked to the development and management of a number of human diseases. Undoubtedly, the gut microbiota has a significant impact on the host immunological response, defense against pathogen overgrowth, biosynthesis, and metabolism. There is a rising need to incorporate strategies for altering the gut microbiota as a means of therapy or infection prevention in routine clinical practice as our understanding of the connections between the gut microbiota and host immunity and infectious illnesses deepens. Probiotics and their derivatives influence the development of various species in the gut microbiome to support the host's health. This review aims to investigate how the gut microbiota is modulated by probiotics, ghost probiotics, postbiotics, and synbiotics, and what this means for infection prevention and antibiotic stewardship.

Keywords: probiotics, ghost probiotics, synbiotics, postbiotics, antimicrobial resistance, antimicrobial stewardship

1. Introduction

The maintenance of a healthy and diverse gut microbiota plays an important role in the prevention and acquisition of antimicrobial resistance, and strategies that modulate its composition have great potential in impacting human health [1]. By limiting the detrimental effects of antimicrobials on the gut flora, the principles of antimicrobial stewardship are strengthened. Antimicrobial stewardship refers to a logical sequence of steps that confirms the effective therapeutic availability of antimicrobials when required [1]. Antimicrobial stewardship has three objectives.

IntechOpen

The first objective is to collaborate with medical professionals to ensure that each patient receives the best antibiotic at the right dosage and for the right amount of time. The correct drug, dose, de-escalation to pathogen-directed therapy, and length of therapy are essential components of the best antimicrobial therapy [2]. Preventing misuse, abuse, and overuse of antibiotics is the second objective. Reduction of the emergence of antimicrobial resistance is the third objective [3].

Antimicrobials are the foundation upon which the health system is standing and are regarded as a global public good that has boosted health care, saved lives, and increased economic advantages [4]. Antimicrobial stewardship optimizes patient antimicrobial functioning to enhance outcomes and lowers antimicrobial resistance. In addition to encouraging the use of agents that are less likely to select for resistant microorganisms, it aids in the development of techniques and interventions that aim to reduce the unnecessary use of antimicrobials. Its initiatives lower prices, antimicrobial resistance, and toxicity [5]. According to Shiffen et al. [6], antimicrobial resistance is the modification of bacteria that renders them resistant to an antimicrobial. Antimicrobial resistance is one of the top ten global public health problems now facing the world [7].

In order to maintain antibiotic effectiveness and prevent the emergence and spread of infections that are antibiotic-resistant, antimicrobial stewardship promotes the cautious application of antimicrobials [8]. Antimicrobial treatments can cause microbes including viruses, fungi, and parasites to adapt and develop resistance. For instance, bacteria that are exposed to antibiotics develop a resistance that allows them to proliferate in the presence of antibiotics and pass on their resistance genes to their progeny [9]. As time goes on, highly resistant strains of microbes develop which are difficult or impossible to treat with available antimicrobials [10]. For antimicrobial therapies to be effective, antimicrobial stewardship programs promote the appropriate use of antibiotics, which includes proper prescription of drugs, the right dose for the right duration, and only when it is necessary [11]. As a result, the selective pressure driving the formation of antimicrobial resistance is reduced. It ensures that these drugs successfully treat infections and reduces the risk of microbial resistance, which is connected to morbidity and mortality [12].

Focus should be placed on the antimicrobial potential of probiotics against pathogenic microbes and host immunity when considering the current pandemic scenario. It has been said that the gut microbiota is an underappreciated organ that constantly establishes bi- or multidirectional interactions with other organs [13]. Therefore, the modification of the human immune system, gut microbiota, and treatment of illness by probiotics and their derivatives has a strong positive impact on reducing antimicrobial resistance and thereby increasing antimicrobial stewardship. The capacity of a healthy microbiota to limit the spread of potentially harmful bacteria is known as colonization resistance [1]. Probiotics play a crucial role in antimicrobial stewardship by maintaining or re-establishing colonization resistance by regulating the gut microbiota to prevent infection [6].

2. Probiotics

Probiotics are living organisms that, when consumed in sufficient proportions, help the host's health, however, the dead bacteria and their components can also exhibit probiotic properties [14]. They help with gastrointestinal health, immune system stimulation, and infection risk reduction [15]. By upsetting the normal equilibrium of gut flora, they reduce antibiotic-associated diarrhea [16]. Additionally, they prevent an infection, which is an objective of antimicrobial stewardship, which reduces the need for antimicrobials. Microbiome therapy aims to modify the gut microbiome by applying native or artificially created bacteria in an additive, subtractive, or modulatory manner. [17]. Probiotic bacteria have the additional capacity to create and release antimicrobial substances like chemical compounds [7].

2.1 Mechanisms of action of probiotics

Probiotics function by:

- Strengthening the epithelial barrier. The gut barrier plays a crucial role in preventing the entry of harmful substances into the bloodstream. Probiotics can enhance the integrity of the intestinal barrier by strengthening the tight junctions between intestinal cells. This helps to maintain a healthy gut lining and reduces the risk of harmful substances crossing into the body [13],

- Boosting adherence to the intestinal mucosa,

- Blocking pathogen adhesion. Probiotics can influence the composition and activity of the gut microbiota. They compete with harmful bacteria for resources and adhesion sites in the gut, effectively crowding out pathogenic microorganisms. Probiotics can also produce antimicrobial substances that inhibit the growth of harmful bacteria [14].

- Eliminating harmful bacteria through competition, creating anti-microbe compounds, and altering the immune system. The colonization of intestinal microbe communities, competitive exclusion of pathogens [18].

- Modulating the immune response. They interact with immune cells in the gut-associated lymphoid tissue (GALT) and stimulate the production of beneficial cytokines, such as interleukin-10 (IL-10), which helps to regulate the immune system and reduce inflammation. This immunomodulatory effect may be beneficial in conditions such as allergies, autoimmune diseases, and certain gastrointestinal disorders [19],

- Production of bioactive compounds: Probiotics can produce bioactive compounds such as vitamins (such as vitamin K and some B vitamins), enzymes, and antimicrobial substances. These compounds can have direct health benefits and contribute to overall production of bacteriocins, modulation of enzymatic activities related to metabolization of specific carcinogens, and production of volatile fatty acids [14]. Probiotics boost the formation of mucin and cell adhesion.

2.2 Effects of probiotics on the gut microbiota

By encouraging the growth of helpful bacteria like *Lactobacillus* and *Bifidobacterium* and suppressing the growth of pathogenic bacteria like *Clostridium difficile*, probiotics can alter the balance of bacteria in the gut [20]. This can help safeguard against certain illnesses while enhancing intestinal health. Depending on the probiotic, harmful bacteria may be killed or its growth may be inhibited [19]. By doing so, the danger of infection can be decreased and pathogenic bacteria in the gut cannot develop [21]. By encouraging the creation of mucus and tight junction proteins, which can help stop harmful bacteria from entering the bloodstream, probiotics can enhance the intestinal barrier [19]. By increasing the production of anti-inflammatory cytokines and decreasing the production of pro-inflammatory cytokines, probiotics can also modify the immune system [20]. This may lessen intestinal inflammation and enhance gut health. A number of infectious and non-infectious diseases can be effectively treated using probiotics and their derivatives.

2.3 Role of probiotics in antimicrobial stewardship

The World Health Organization [11] has come to the conclusion that probiotics are the second-most important immune defense mechanism when frequently advised antimicrobials are unsuccessful against a limited number of pathogens. A number of beneficial intestinal bacteria, including *Lactobacillus*, *Bifidobacterium*, and *Enterococcus*, as shown in **Table 1**, contribute to the improvement of gut microbiota stability [28]. Probiotics can be found in food, beverages, and dietary supplements. Additionally, consuming fermented foods, which are probiotic bacteria's natural habitat, is of immense importance which eventually helps in the better treatment of several diseases, including gut-related disorders [28].

Probiotic	Reported effects	References
Lactobacillus rhamnosus GG	Improved gut health, reduced diarrhea	[22]
Bifidobacterium animalis DN-173010	Enhanced immune response - They can boost antibody synthesis, amplify immune cell activity, and fortify the intestinal barrier, lowering the risk of opportunistic infections while using antibiotics; fewer stomach aches and pains	[23]
Saccharomyces boulardii	AAD (antibiotic-associated diarrhea) prevention and treatment. The natural balance of benevolent bacteria in the stomach can be upset by antibiotics, which can result in diseases like AAD. When given simultaneously with antibiotics, probiotics can aid in microbial restoration and lower the risk of AAD.	[22]
Lactobacillus acidophilus NCFM	Reduction of lactose intolerance	[24]
Streptococcus thermophilus	Enhanced digestion of lactose	[22]
Lactobacillus plantarum 299v	Improved symptoms of irritable bowel syndrome	[25]
Bifidobacterium bifidum	Enhanced immune function	[25]
Lactobacillus *Streptococcus*	Treatment Adherence - Probiotics may improve a patient's treatment adherence with antibiotics. Probiotics have been shown to improve treatment compliance, decrease side effects, and improve patient satisfaction when added to antibiotic regimens.	[26]
Lactobacillus *Saccharomyces*	Inhibiting the growth and colonization of *Clostridium difficile*	[27]

Table 1.
Roles of probiotics.

If it happens that once in the human body the probiotics die, they can no longer have a positive effect on health therefore scientists are conducting research to formulate technologies that produce dead and broken-down bacteria which are like probiotics or better in terms of immune enhancing properties [29].

3. Ghost probiotics

It has recently been discovered that ghost probiotics, or inactivated or heat-killed probiotic cells, can be used in place of live probiotics. In both animals and people, they have demonstrated effectiveness in modifying anti-inflammatory and pro-inflammatory immune responses [30]. Ghost probiotics, commonly referred to as para-probiotics, are beneficial microorganisms that have been rendered inactive. They can either be whole or broken [8]. Provoking the human immune system, promoting a favorable immunological response, and having anti-inflammatory effects in both people and animals are all impacts of ghost probiotics [31]. They are favored because they have a longer shelf life, which makes it easier to distribute them without refrigeration to various locations. In immune-compromised people, the use of ghost probiotics lowers the likelihood of contamination, microbial translocation, or heightened inflammatory immune responses [8, 32].

3.1 Mechanisms of action of ghost probiotics

Probiotics that have been inactivated may influence immunological responses by interacting with the immune system. A study suggested that the beneficial effects of ghost probiotics on gut microbiota may be due to the release of soluble factors such as cell membrane fragments and metabolites that can modulate immune function and promote the growth of beneficial bacteria [4]. These immuno-modulatory actions might suppress immune-related disorders and control immune response and inflammation. In the stomach, they fight against harmful microbes for adhesion sites and nutrition. They might hinder the attachment and proliferation of harmful bacteria by occupying these locations. In another study, researchers discovered that supplementation with ghost probiotics derived from *Lactobacillus plantarum* had positive effects on gut microbiota, including an increase in beneficial bacteria such as *Bifidobacterium* and *Lactobacillus* and a decrease in potentially harmful bacteria such as *Clostridium* [17].

They might help keep the intestinal barrier healthy and functioning properly. They are able to interact with gut epithelial cells, enhancing tight junction protein production and preserving the integrity of the gut barrier. Inactivated probiotics can aid in preventing the transfer of hazardous chemicals from the gut into the circulation by enhancing the barrier function [33]. Even in their non-viable state, inactivated probiotics may retain some antimicrobial properties. They may produce antimicrobial compounds such as organic acids, bacteriocins, or antimicrobial peptides that can inhibit the growth of pathogenic bacteria in the gut. These antimicrobial effects contribute to maintaining a balanced gut microbiota [8]. Inactivated probiotics may still have metabolic effects. They can interact with dietary components and release bioactive compounds such as short-chain fatty acids (SCFAs) or other metabolites. SCFAs, for example, can provide energy to colon cells, support the gut barrier function, and have systemic effects on metabolism [32].

Ghost probiotic	Reported effects	References
Non-pathogenic *Escherichia coli* Nissle 1917	Competitive exclusion of pathogenic bacteria. They compete with pathogenic bacteria for binding sites and nutrients in the gut. This competitive exclusion mechanism can help prevent the attachment and colonization of harmful bacteria, reducing the risk of infections.	[26]
Lactobacillus rhamnosus GG	Reduced inflammatory response in the gut. When it comes to viral disorders, where inflammation can cause tissue damage, they can help regulate immune responses and minimize excessive inflammation.	[35]
Bacillus subtilis	Enhanced immune response; inhibition of pathogens	[36]
Lactobacillus acidophilus	Modulation of immune responses. Similar to living probiotics, they influence the immune system by promoting the synthesis of substances that modulate the immune system. They have the power to affect immune cell activity and cytokine synthesis, which can strengthen immunological responses and promote the body's fight against infections.	[37]
Bifidobacterium	Boost intestinal health and aid in keeping the intestinal barrier's integrity. They can strengthen tight connections between cells, increase mucus production, and improve the gut epithelium's defensive capabilities. This may help lower the possibility of bacterial translocation and ensuing illnesses.	[24]
Streptococcus	Treat and prevent infections	[38]
Lactobacillus *Streptococcus* *Bifidobacterium*	Reduce use of antibiotics. In some circumstances, they might act as an additional therapy to lessen the requirement for antibiotics. Ghost probiotics may enhance the body's natural defense mechanisms by boosting the immune system and encouraging a healthy gut flora, potentially lowering the need for antibiotics and limiting the emergence of antibiotic resistance.	[39]

Table 2.
Roles of ghost probiotics.

3.2 Roles of ghost probiotics in antimicrobial stewardship

In antimicrobial stewardship, ghost probiotics are used as a possible remedy in fighting against antimicrobial resistance [34], as illustrated in **Table 2**. A study by Thursby and Juge [40], revealed a significant decrease in the number of infections in the treatment group compared to the control group when the efficacy of a ghost probiotic formula in preventing respiratory tract infections in a population of elderly individuals in a nursing home was tested. In 2019, another study investigated the potential use of ghost probiotics as a preventive strategy against *Clostridium difficile* infections in a mouse model. Researchers found that a ghost probiotic formula of *Lactobacillus reuteri* DSM 17938 significantly shortens the duration of acute infectious diarrhea and reduces abdominal pain in patients with colitis [19, 41].

4. Synbiotics

The term synbiotics describes probiotics and prebiotics that have been combined in dietary supplements or food ingredients to create a synergistic effect. According to Wang et al. [42], prebiotics are common foods that strengthen the immune system while encouraging the development of probiotic bacteria in the human stomach.

Prebiotics, which are necessary for the development and function of microorganisms, should not be mistaken with probiotics. Supplements and specific foods like yogurt, whole grains, and sauerkraut contain synbiotics [43]. *Bifidobacterium*, *Lactobacillus*, *Saccharomyces boulardii*, *Bacillus cougulans*, and prebiotics like Inulin and XOS are the most widely utilized synbiotics [44]. The most popular fibers that are typically utilized in conjunction with probiotics are fructans. Synbiotics were created in order to get over potential probiotic survival issues [43]. Synbiotics play a significant role in gut bacterial balance, which benefits the human immune system, metabolism, and gut health [43].

4.1 Effects of synbiotics on gut microbiota

The gut microbiota appears to play a role in the pathogenesis of obesity and associated diseases. Therefore, gut microbiota can also be considered a promising target in the comprehensive dietary approach to the prevention and treatment of obesity, including weight loss and weight maintenance [45]. Synbiotic supplementation increases the abundance of gut bacteria associated with positive health effects, especially *Bifidobacterium* and *Lactobacillus*, and it also appears to increase the gut microbiota richness [45].

According to Roy and Dhaneshwar [46], synbiotics can aid in enhancing the integrity of the intestinal barrier, which may lessen intestinal inflammation and enhance immunological function. Short chain fatty acids (SCFAs) are vital energy sources for colonocytes and have anti-inflammatory characteristics. Synbiotics help to boost production of SCFAs [47]. According to Phavichitr et al. [45], synbiotics can determine the abundance of harmful bacteria in the stomach, including *Escherichia coli* and *Clostridium difficile*. Additionally, synbiotics may enhance cholesterol, glucose, and inflammation metabolism, which could be advantageous for metabolic health [44].

4.2 How do synbiotics work?

Synbiotics work by combining prebiotics and probiotics to provide a synergistic effect that promotes optimal gut health. The probiotic adds beneficial bacteria that greatly benefit the gut microbiome. The prebiotic fuels the beneficial bacteria that is already present in the gut microbiome. There are two types of synbiotics which are synergistic synobiotics and complementary synbiotics [43]. Synergistic synbiotics are formulated to have live probiotics fueled by the prebiotic substrate that is co-administered in the product, allowing the two elements to work together as a self-contained whole [44]. Complementary synbiotics are formulated so that both elements work independently, with the prebiotic chosen to target resident microorganisms in the gut [44].

Synbiotics help to promote the growth of beneficial bacteria in the gut, leading to improved gut microbiome balance and overall gut health. A study by Palai *et al.* [43], supported that symbiotic supplements used in clinical trial modulate human gut microbiota by increasing abundance of potentially beneficial microbial species. Synbiotics have been shown to boost immune function by promoting the growth of beneficial bacteria that can help fight off harmful pathogens. A study showed that a synbiotic consisting of *Lactobacillus acidophilus* and Inulin improved the immune response in elderly individuals by increasing the production of immunoglobulin A and reducing the levels of pro-inflammatory cytokines such as interleukin-6 and tumor necrosis factor-alpha [44].

Synbiotics can improve digestion by promoting the growth of beneficial bacteria that aid in the breakdown and absorption of nutrients. They display positive effects on digestion and enzymatic activity, which may help alleviate many of the symptoms associated with lactose intolerance. For instance, the synbiotic of *Lactobacillius acidophilus* and Inulin improves the digestion of lactose in lactose-intolerant individuals by increasing the population of lactase-producing bacteria in the gut [47]. Also, probiotics like *Bifidobacterium lactis* combined together with fructooligosaccharides (FOS) improve the digestion of protein and fat in healthy adults by enhancing the production of pancreatic enzymes. Synbiotics have been shown to have anti-inflammatory effects, which can help to reduce inflammation in the gut and throughout the body. For example, research showed that a synbiotic consisting of *Lactobacillus acidophilus* and FOS reduces levels of C-reactive protein, a cause of inflammation [48].

4.3 Roles of synbiotics in antimicrobial stewardship

Synbiotics play a valuable role in antimicrobial stewardship by reducing the need for antimicrobial agents and preventing the development of antibiotic resistance. Synbiotics can be used to prevent and treat a number of conditions that are often treated with antimicrobials, such as *Clostridium difficile* infection, inflammatory bowel disease, and irritable bowel syndrome [49]. *Lactobacillus acidophilus* and *Bifidobacterium lactis* with FOS have been shown to reduce the incidence of antibiotic-associated diarrhea and *Clostridium difficile* infection in hospitalized patients efficiently [47]. *Lactobacillus rhamnosus* GG and inulin have been shown to reduce the incidence of antibiotic-associated diarrhea in children [47]. *Akkermansia muciniphila* and prebiotic fibers have been shown to improve gut barrier function and reduce inflammation in obese and overweight adults [49].

5. Postbiotics

Postbiotics are metabolic by-products of probiotic bacteria [34]. These byproducts include short-chain fatty acids, enzymes, peptides, and other compounds that can have various health benefits. Postbiotics play a role in maintaining gut health, regulating the immune system, providing anti-inflammatory effects and potentially reducing the risk of certain diseases [2]. Postbiotics can be naturally produced by the probiotic bacteria that live in our gut and they can also be produced through fermentation of certain foods such as yogurt, kefir, sauerkraut, and kimchi [50].

5.1 Roles of postbiotics in antimicrobial stewardship

Postbiotics, such as peptides and organic acids, have been shown to have beneficial effects on gut microbiota as shown in **Table 3**. Certain peptides produced by *Lactobacillus* species have been shown to inhibit the growth of pathogenic bacteria such as *Escherichia coli* and *Salmonella enteritidis* [9]. Organic acids such as acetic acid and lactic acid, which are produced by some probiotic bacteria, can help to lower the pH of the gut, creating an environment that is less favorable for the growth of harmful bacteria [2]. Short-chain fatty acids (SCFAs) are produced by the fermentation of dietary fiber by gut bacteria and have been shown to have antimicrobial properties. They can help to inhibit the growth of harmful bacteria such as *Clostridium difficile*, which is a common cause of antibiotic-associated diarrhea. By promoting the growth of beneficial

Postbiotic	Reported effects	References
Short chain fatty acids from gut microbiota fermentation	Enhanced gut barrier function; anti-inflammatory effects. They can reduce the production of pro-inflammatory molecules, promote the production of anti-inflammatory substances, and help alleviate inflammation associated with infections.	[51]
Lactic acid bacteria metabolites	Modulation of immune responses; antimicrobial properties	[52]
Bacteriocins	Inhibition of pathogenic bacteria	[53]
Extracellular vesicles	Cellular communication; immune regulation	[54]
Organic acids	pH regulation; antimicrobial effects	[55]
Peptidoglycan derived muropeptides	The intestinal barrier's integrity and functionality can be supported by immunomodulation. They can improve the protective qualities of the barrier, encourage mucus secretion, and fortify the tight junctions between intestinal cells. This aids in stopping dangerous bacteria and poisons from crossing the intestinal barrier.	[51]
	They can aid in forming and preserving a balanced makeup of the gut flora. They can support a healthy microbial environment by selectively promoting the development of advantageous bacteria, preventing the growth of diseases, and both. This may improve infection resistance and intestinal health.	[56]
	Reduction of negative Effects Related to Antibiotic Use - They might help to lessen the negative effects related to antibiotic use. They can aid in reducing antibiotic-associated diarrhea, maintain gut health while taking antibiotics, and hasten recovery by encouraging a balanced gut flora.	[35]

Table 3.
Roles of postbiotics.

bacteria and reducing the need for antibiotics, SCFAs can help to reduce the overall use of antimicrobials and limit the development of antimicrobial resistance [50].

Bacteriocins are antimicrobial peptides produced by some probiotic bacteria that can help to inhibit the growth of pathogenic bacteria. For example, nisin, a bacteriocin produced by *Lactococcus lactis*, has been shown to be effective against a range of Gram-positive bacteria, including *Staphylococcus aureus* and *Streptococcus pneumoniae* [57]. By using bacteriocins instead of antimicrobials to target specific pathogens, postbiotics can help to reduce the overall use of antimicrobials and limit the development of antimicrobial resistance.

Exopolysaccharides are complex carbohydrates produced by some probiotic bacteria that can have prebiotic effects, meaning they can promote the growth of beneficial bacteria in the gut [50]. Peptidoglycans are complex molecules found in the cell walls of bacteria that can modulate the immune system and have anti-inflammatory effects. Cell surface proteins are proteins found on the surface of probiotic bacteria that can interact with host cells and modulate immune responses [33].

6. Issues surrounding the use of probiotics and their derivatives in antimicrobial stewardship

To be considered a probiotic, a supplement needs to meet all the requirements set forth by the 2002 Food and Agriculture Organization/World Health [57].

The requirements stipulate that a probiotic must be a live microorganism, needs to be administered and should have a health benefit. Additionally, a probiotic must be safe and void of vectors that can transfer resistance to antibiotics [14]. One issue surrounding the use of probiotics in antimicrobial stewardship is the potential for interactions between probiotics and antibiotics. Some studies have suggested that probiotics may reduce the effectiveness of antibiotics by competing with them for absorption in the gut or by producing antibacterial substances that can neutralize antibiotics. This could lead to suboptimal treatment outcomes and the emergence of antibiotic-resistant strains of bacteria [58]. However, other studies have shown that probiotics can enhance the effectiveness of antibiotics by improving the gut microbiota and reducing the risk of antibiotic-associated diarrhea and other side effects. Another issue is the lack of standardization in the production and labeling of probiotics [57]. There is significant variability in the quality and efficacy of probiotic products, and there are currently no widely accepted standards for the selection, testing, and labeling of probiotics. This makes it difficult for healthcare providers to make informed decisions about which probiotics to use and for what indications [59].

7. Conclusion and future perspectives

In the midst of the ongoing antibiotic resistance challenge, using probiotics, postbiotics, ghost probiotics, and synbiotics offers a secure alternative to treating microbial infections. In order to research future potential uses of probiotics in both human and animal diseases, it is vital to update our knowledge in this area. As a result, this work may be used as a reference to understand naturally occurring probiotic compounds and their prospective applications for the treatment and management of numerous human diseases, providing a significant contribution to antimicrobial stewardship. Probiotics and their derivatives that exhibit powerful activity, both alone and in combination, must therefore be researched in the future.

Author details

Nesisa Nyathi[1*], Duduzile Ndhlovu[1], Esma Rabvukwa[1] and Abigarl Ndudzo[1,2]

1 Department of Applied Biotechnology, Lupane State University, Zimbabwe

2 Department of Molecular Biology and Biotechnology, Pan African University of Basic Sciences and Technology, Jomo Kenyatta University of Agriculture and Technology, Juja, Kenya

*Address all correspondence to: nesisamdluli@gmail.com

References

[1] Zawistowska-Rojek A, Tyski S. Are probiotic really safe for humans? Polish Journal of Microbiology. 2018;**67**(3): 251-258. DOI: 10.21307/pjm-2018-044

[2] Aguilar-Toala JE, Garcia-Varela R, Garcia HS, Mata- Haro V, Gonzalez-Cordova AF, Vallejo-Cordoba B. Postbiotics: An evolving term within the functional foods field. Trends Food Science and Technology. 2018;**75**:105-114. DOI: 10.1016/j.tifs.2018.03.009

[3] Jeong K, Kim M, Jeon SA, Kim YH, Lee S. A randomized trial of *lactobacillus rhamnosus* IDCC 3201 tyndallizate (RHT3201) for treating atopic dermatitis. Pediatric Allergy and Immunology. 2020;**31**:783-792. DOI: 10.1111/pai.13269

[4] Arboleya S, Merinero MC, Solis G. Ghost probiotics with a combined regimen: A novel approach for the prevention and treatment of neonatal sepsis. Beneficial Microbes. 2021;**12**(1). DOI: 10.3920/BM2020.0079

[5] Hoarau C, Martin L, Faugaret D, Baron C, Dauba A, Aubert-Jacquin C, et al. Supernatant from Bifidobacterium differentially modulates transduction signaling pathways for biological functions of human dendritic cells. PLoS One. 2018;**3**(7):E2753. DOI: 10.1371/journal.pone.0002753

[6] Sniffen JC, McFarland LV, Evans CT, Goldstein EJC. Choosing appropriate probiotic product for your patient: An evidence based particular guide. PLoS One. 2018;**13**(12):e0202205. DOI: 10.1371/journal.pone.0209205

[7] Szczerbiec D, Piechoka J, Glowacki R, Torzewska A. Organic acids secreted by lactobacillus spp. isolated from urine and their antimicrobial activity against Uropathogenic *Proteus mirabilis*. Molecules. 2022;**27**:5557. DOI: 10.3390/molecules27175557

[8] Kim WK, Jang YJ, Han DH, Jeon K, Lee C, Han HS. *Lactobacillus paracasei* KBL382 administration attenuates atopic dermatitis by modulating immune response and gut microbiota. Gut Microbes. 2020;**12**:1-14. DOI: 10.1080/19490976.2020.1819156

[9] Chang CM, Tsai MH, Liao WC, Yang PH, Li SW, Chu SM, et al. Effects of probiotics on gut microbiomes of extremely preterm infants in the neonatal intensive care unit: A prospective cohort study. Nutrients. 2022;**14**(15):3239. DOI: 10.3390/nu14153239

[10] Sun M, Luo J, Liu H, Xi Y, Lin Q. Can mixed strains of *lactobacillus* and *Bifidobacterium* reduce eczema in infants under three years of age? A meta-analysis. Nutrients. 2021;**13**. DOI: 10.3390/nu13051461

[11] World Health Organisation. Antimicrobial Stewardship Programmes in Health Care Facilitating Low- and Middle Income Countries: A WHO Practical Toolkit. 2019. Available from: https://apps.who.int/iris/bitstream/handle/10665/329404/9789241515481-eng.pdf [Accessed: 20 May 2023]

[12] Azim-Majumder MA, Rahman S, Cohall D, Bharatha A, Singh K, Haque M, et al. Antimicrobial stewardship: Fighting antimicrobial resistance and protecting global public health. Infection and Drug Resistance. 2020;**13**:4713-4738. DOI: 10.2147/IDR.S290835

[13] Halluran K, Underwood MA. Probiotics mechanisms of action. Early Human Development. 2019;**135**:58-65. DOI: 10.1016/j.earlhumdev.2019.05.010

[14] Plaza-Diaz J, Ruiz-Ojeda FJ, Gil-Campos M, Gil A. Mechanisms of action of probiotics. Advanced Nutrition. 2019;**10**. DOI: 10.1093/advances/nmy063

[15] Reid G, Gadir AA, Dhir R. Probiotics: Reiterating what they are and what they are not. Frontiers in Microbiology. 2019;**10**. DOI: 10.3389/fmicb.2019.00424

[16] Blaabjerg S, Artzi DM, Aabenhus R. Probiotics for the prevention of antibiotic-associated diarrhea in outpatients- a systematic review and meta-analysis. Antibiotics. 2017;**6**(4). DOI: 10.3390/antibiotics6040021

[17] Yadav M, Chauhan NS. Microbiome therapeutics: Exploring the present scenario and challenges. Gastroenterology Report. 2021;**10**:goab046. DOI: 10.1093/gastro/goab046

[18] Li Y, Xia S, Jiang X, Feng C, Gong S, Ma J, et al. Gut microbiota and diarrhea: An updated review. Frontiers in Cellular and Infection Microbiology. 2021;**11**. DOI: 10.3389/fcimb.2021.625210

[19] Wu J, Zhang Y, Ye L, Wang C. The anti-cancer effects and mechanisms of lactic acid bacteria exopolysaccharides in vitro: A review. Carbohydrate Polymers. 2021;**253**:117308. DOI: 10.1016/j.carbpol.2020.117308

[20] Divakar D, Poonam SN. The gut microbiota influenced by the intake of probiotics and functional foods with prebiotics can sustain wellness and alleviate certain ailments like gut-inflammation and colon cancer. 2022. DOI: 10.3390/microorganisms10030665

[21] Case I. The Regulation of Probiotics. Degree Candidate for Masters of Science in Regulatory Science. Johns Hopkins University.Available from: https://www.jhsph.edu/research/centers-and-institutes/center-of-excellence-in-regulatory-science-and-innovation/training/Iris; 2021 [Accessed 24 May 2023]

[22] George Kerry R, Patra JK, Gouda S, Park Y, Shin H, Das G. Benefaction of probiotics for human health: A review. Journal of Food and Drug Analysis. 2018;**26**(3):927-939. DOI: 10.1016/j.jfda.2018.01.002

[23] Chen M, Lin W, Li N, Wang Q, Zhu S, Zeng A, et al. Therapeutic approaches to colorectal cancer via strategies based on modulation of gut microbiota. Frontiers in Microbiology. 2022;**13**:945533. DOI: 10.3389/fmicb.2022.945533

[24] Bozomitu L, Miron I, Adam Raileanu A, Lupu A, Paduraru G, Marcu FM, et al. The gut microbiome and its implication in the mucosal digestive disorders. Biomedicine. 2022;**10**(12):3117. DOI: 10.3390/biomedicines10123117

[25] Mazziotta C, Tognon M, Martini F, Torreggiani E, Rotondo JC. Probiotics mechanism of action on immune cells and beneficial effects on human health. Cell. 2023;**12**(1). DOI: 10.3390/cells12010184

[26] Aponte M, Murru N, Shoukat M. Therapeutic, prophylactic, and functional use of probiotics: A current perspective. Frontiers in Microbiology. 2020;**11**:562048. DOI: 10.3389/fmicb.2020.562048

[27] Gunaratnam S, Millette M, McFarland LV, DuPont HL, Lacroix M. Potential role of probiotics in reducing Clostridioides difficile virulence:

Interference with quorum sensing systems. Microbial Pathogenesis. 2021;**153**:104798. DOI: 10.1016/j.micpath.2021.104798

[28] Prakoeswa C, Bonita L, Karim A, Herwanto N, Umborowat MA, Setyaningrum T. Beneficial effect of *lactobacillus plantarum* IS-10506 supplementation in adults with atopic dermatitis: A randomized controlled trial. Journal of Dermatological Treatment. 2022;**33**:1491-1498. DOI: 10.1080/09546634.2020.1836310

[29] Marco ML, Heeney D, Binda S, Cifelli CJ, Cotter PD, Foligne B, et al. Health benefits of fermented foods: Microbiota and beyond. Current Opinions in Biotechnology. 2017;**44**:94-102. DOI: 10.1016/j.copbio.2016.11.010

[30] Siciliano RA, Reale A, Mazzeo MF, Morandi S, Silvetti T, Brasca M. Paraprobiotics: A new perspective for functional foods and nutraceuticals. Nutrients. 2021;**13**(4). DOI: 10.3390/nu13041225

[31] Hasan A, Paray BA, Hussain A, Qadir FA, Attar C, Aziz FM. A review on the cleavage priming of the spike protein on coronavirus by angiotensin-converting enzyme-2 and furin. Journal of Biomolecular Structure and Dynamics. 2021;**39**(8):3025-3033. DOI: 10.1080/07391102.2020.1754293

[32] Ndudzo A, Ndlovu S, Nyathi N, Makuvise-Sibanda A. Unlocking the potential of ghost probiotics in combating anti-microbial resistance. In: The Global Antimicrobial Resistance Epidemic: Innovative Approaches and Cutting Edge Solutions. London, UK: IntechOpen; 2022. DOI: 10.5772/intechopen.10426

[33] Kim D, Kim YJ, Kim SH, Yang JS. Short-chain fatty acids produced by *lactobacillus plantarun* strain Lp27 inhibit *Escherichia coli* 0157:H7 and *salmonella Typhimurium* in vitro. Journal of Microbiology and Biotechnology. 2019;**29**(8):1195-1205

[34] Aguilar-Toalá JE, Cuevas-Gonzalez P, Liceaga A. Postbiotics and Parabiotics: From concepts to applications. Food Research International. 2020;**136**:109502. DOI: 10.1016/j.foodres.2020.109502

[35] Amara AA, Shibl A. Role of probiotics in health improvement, infection control and disease treatment and management. Saudi Pharmaceutical Journal: SPJ. 2015;**23**(2):107-114. DOI: 10.1016/j.jsps.2013.07.001

[36] Incrocci R, Negris O, McGrath S, Swartzendruber JA. Bacillus subtilis provides long-term protection in a murine model of allergic lung disease by influencing bacterial composition. Allergie. 2023;**3**(1):1-10. DOI: 10.3390/allergies3010001

[37] Li HY, Zhou DD, Ren Y, Huang S, Zhao C, Shang A, et al. Effects and mechanisms of probiotics, prebiotics, synbiotics and post biotics on metabolic diseases targeting. Gut Microbiota. 2021;**13**(9):3211. DOI: 10.3390/nu13093211

[38] Kechagia M, Basoulis D, Konstantopoulou S, Dimitriadi D, Gyftopoulou K, Skarmoutsou N, et al. Health benefits of probiotics: A review. ISRN Nutrition. 2013. DOI: 10.5402/2013/481651

[39] Patangia DV, Ryan CA, Dempsey E, Ross RP, Stanton C. Impact of antibiotics on the human microbiome and consequences for host health. Microbiology Open. 2022;**11**(1). DOI: 10.1002/mbo3.1260

[40] Thursby E, Juge N. Introduction to the human gut microbiota. The

Biochemical Journal. 2017;**474**(11):1823-1836. DOI: 10.1042/BCJ20160510

[41] Saviano A, Brigida M, Migneco A, Gunawardena G, Zanza C, Candelli M, et al. *Lactobacillus Reuteri* DSM 17938 (*Limosilactobacillus reuteri*) in diarrhea and constipation: Two sides of the same coin? Mecina (Kaunas). 2021;**57**(7):643. DOI: 10.3390/medicina57070643

[42] Wang HT, Anvari S, Anagnostou K. The role of probiotics in preventing allergic disease. Children (Basel, Switzerland). 2019;**6**(2):24. DOI: 10.3390/children6020024

[43] Palai S, Derecho CMP, Kesh SS, Egbuna C, Onyeike PC. Prebiotics, probiotics, synbiotics and its importance in management of diseases. Chapter 10. In: Functional Foods and Neutraceuticals. Springer; 2020. DOI: 10.1007/978-3-030-42319-3_10

[44] Thilagavathi T. Probiotics, prebiotics, synbiotics and its health benefits, International Journal of Current Microbiology and Applied Sciences. 2020 DOI: 10.20546

[45] Phavichitr N, Wang S, Chomoto S, Tantibhaedhyangkul R, Kakourou A, Intarakhao S, et al. Impact of Synbiotics on gut microbiota during early life: A randomized double-blind study. Scientific Reports. 2021;**11**(1):3534. DOI: 10.1038/s41598-021-83009-2

[46] Roy S, Dhaneshwar S. Role of prebiotics, probiotics, and synbiotics in management of inflammatory bowel disease: Current perspectives. World Journal of Gastroenterology. 2023;**29**(14):2078-2100. DOI: 10.3748/wjg.v29.i14.2078

[47] Chatchatee P, Way SL, Eugenia C, Kosuwon P, Simaakachorn N, Yavuz Y, et al. Effects of growing up supplemented with prebiotics and LCPUFAS on infectious in young children. 2018. DOI: 10.1097/MPG.0000000000000252

[48] Xie A, Chen A, Chen Y, Luo Z, Jiang S, Chen D, et al. *Lactobacillus* for the treatment and prevention of atopic dermatitis: Clinical and experimental evidence. Frontiers in Cellular and Infection Microbiology. 2023;**13**:1137275. DOI: 10.3389/fcimb.2023.1137275

[49] Depommier C, Everard A, Druart C, Plovier H, Van Hul M, Vieira-Silva S, et al. Supplementation with Akkermansia muciniphila in overweight and obese human volunteers: A proof-of-concept exploratory study. Nature Medicine. 2019;**25**(7):1096-1103. DOI: 10.1038/s41591-019-0495-2

[50] Zhang L, Chen H, Wang Y, Yu L. Antimicrobial activity of acetic acid produced by lactobacillus plantarum JD-15 against foodborne pathogens. Journal of Agricultural and Food Chemistry. 2019;**67**(43):12059-12067. DOI: 10.3390%2Ffoods10123131

[51] Delgado S, Sánchez B, Margolles A, Ruas-Madiedo P, Ruiz L. Molecules produced by probiotics and intestinal microorganisms with immunomodulatory activity. Nutrients. 2020;**12**(2). DOI: 10.3390/nu12020391

[52] Moon A, Sun Y, Wang Y, Huang J, Zafar Khan MU, Qiu H. Lactic acid bacteria as mucosal immunity enhancers and antivirals through Oral delivery. Applied Microbiology. 2022;**2**(4):837-854. DOI: 10.3390/applmicrobiol2040064

[53] Francisco D, Alexis J, Rubén J. Bacteriocins: An overview of antimicrobial, toxicity, and biosafety assessment by in vivo models. Frontiers in Microbiology. 2021;**12**:630695. DOI: 10.3389/fmicb.2021.630695

[54] Buzas EI. The roles of extracellular vesicles in the immune system. Nature Reviews Immunology. 2023;**23**(4):236-250. DOI: 10.1038/s41577-022-00763-8

[55] Ma L, Tu H, Chen T. Postbiotics in human health: A narrative review. Nutrients. 2023;**15**(2). DOI: 10.3390/nu15020291

[56] Hou K, Wu Z, Chen X, Wang J, Zhang D, Xiao C, et al. Microbiota in health and diseases. Signal Transduction and Targeted Therapy. 2022;**7**(1):1-28. DOI: 10.1038/s41392-022-00974-4

[57] FDA. 2017. Food Labeling: Revision of the Nutrition and Supplement Facts Label. Available form: https://www.regulations.gov/document/FDA-2012-N-1210-0875 [Accessed 19 May 2023]

[58] de Simone C. The unregulated probiotic market. Clinical Gastroenterology and Hepatology. 2019;**17**(5):809-817. DOI: 10.1016/j.cgh.2018.01.018

[59] Grebow J. Will FDA Ever Provide a Practical Regulatory Path for Probiotics? Panel Discusses at NPS's the Big Natural Conference. 2020. Available from: https://www.nutritionaloutlook.com/view/will-fda-ever-provide-a-practical-regulatory-path-forprobiotics-panel-discusses-at-npa-s-the-big-natural-conference [Accessed: 19 May 2023]

Chapter 7

Laboratory Information Management System (LIMS) Analysis: A Lucrative Tool to Support the Control of Antimicrobial Resistance (AMR)

Maryam Nasrumminallah and Fatima Rehan

Abstract

Globally, antimicrobial resistance (AMR) poses a real risk to people's health. To ascertain the burden, implications, and trends of AMR and to track the results of interventions, surveillance is a crucial activity. High-quality laboratory data must be efficiently collected and shared with surveillance systems. Particularly in LMICs with high disease loads, the capacity of laboratories is being dramatically increased. Building capacity for effective laboratory data administration is still a resource-constrained issue that, if left unresolved, may hamper development toward comprehensive AMR surveillance in LMICs. The absence of an open-source, useful lab tests database is particularly concerning. In this Personal View, we present an overview of the laboratory data management practices in Lower Middle Income Countries (LMIC) laboratories, a snapshot of the technological requirements for microbiological lab data management, and a description of the critical remedial measures. Current efforts to boost capability for AMR surveillance in LMICs would not be entirely successful without strategies for upgrading information technology equipment and information management systems in microbial labs.

Keywords: laboratory information management, AMR, bacterial antigen, antibiotics, surveillance resistance, patient care

1. Introduction

Worldwide, antibiotic use has spared millions of souls. However, overuse of antibiotics over the past 60 years has resulted in the evolution of antibiotic-resistant bacteria, hastening the spread of fatal infectious diseases and burdening society with financial costs. Even though this issue is frequently thought of as a worldwide one, some parts of the world are ignorant of its true extent [1]. AMR, an evolving danger to the general population, is caused by resistant strains of antibiotics. is making it harder to treat and prevent bacterial infections with today's drugs.

Antibiotic-resistant organisms have emerged due to widespread antibiotic use during the past 60 years, accelerating the spread of fatal diseases and adding to the financial burden on society. There is no denying that this problem exists on a worldwide scale, yet in some regions of the world its scope is not quite evident [1].

The tendency of microorganisms to resist treatment or control via antimicrobial drugs adopted to do so is known as antibiotic resistance. Antimicrobial resistance (AMR) could pose a serious worldwide issue. Information on the rate of bacteria immune to antibiotics in countries of low- and middle socioeconomic status is scarce. Antibiotic resistance, however, is a significant international and domestic problem in West Pakistan, one of South Asia's growing nations [2]. Recent discoveries of MDR (multi-drug resistant) and highly drug resistant microbes have both been made in Pakistan. The enterobacteriaceae family has developed quinolone resistance in Pakistan [3]. In addition, 93.7% of strains were found to be resistant to third-generation cephalosporins in a research on blood stream infections (BSI) [4]. The theory, which is commonly backed by a number of studies, contends that patients' inaccurate indefinite consumption is to fault, along with irrational advice, incentives for over-prescribing, self-treatment, less trained personnel, and a lack of professional training and standard examinations.

Suboptimal antimicrobial use can be caused by a number of things, including a general overreliance on empirical antimicrobial therapy with a concomitant disregard for microbiological outcomes, errors in testing at the lab, inability to bring in suitable samples for culture, unauthorized use of microbes in order assets and the absence of a microbiologically verified evaluation.

1.1 What is an antimicrobial stewardship program?

The Infectious Diseases Society of America (IDSA), the Society for Healthcare Epidemiology of America (SHEA), and the pediatric Infectious Diseases Society (PIDS), all agree that antibiotic stewardship involves collaborative efforts to encourage the selection of the best medication regimen, including the administration, time span, and route of administration. The purpose of these interventions is to measure and enhance the proper use of antibiotics [5].

The implementation of evidence-based treatments that maximize the use of antibiotics while limiting the emergence of resistance is the focus of ASPs. These programs include the creation of guidelines as well as regulatory and policy measures. Practically speaking, ASPs could be considered as the right course of action for employing antibiotics in a way that provides ongoing access to efficient treatment while minimizing side effects [5].

1.2 What is needed to set up an AMR surveillance system?

AMR surveillance typically requires a supportive environment and a commitment to delivering high-quality care in order for specialists from various facets of the medical field to follow good clinical practice, connect effectively, coordinate customs, and complete projects in the shortest possible time. Based on an effective infectious disease diagnostic cycle, this multifaceted or collaborative strategy includes clinicians who collect clinical samples for the microbiology lab and submit them there, a bacteriology lab that can perform species recognition and AST, as well as a system to report, collect, analyze, and analyze facts so that those who must take action are informed [6]. The uniform and verified information gathered from an effective

AMR surveillance system can help countries create practical therapy recommendations, based on research public health policies, and other actions. There are a few key components for creating an AMR surveillance system. The system can grow gradually once these foundational pieces are in place [6].

1.3 Justification for surveillance of antimicrobial resistance

Data from AMR surveillance programs are used to track how bacteria respond to various antibiotic agents. Regular data distribution can aid in the revision of case management guidelines for healthcare facilities and aid in the methodical eradication of AMR. Additionally, this information can be utilized to educate the public, regulators, pharmacists, and medical professionals. To prove the effectiveness of treatment when treating communities during outbreaks, AMR surveillance is essential. It is crucial for both identifying the formation of novel resistance patterns and tracking the effectiveness of measures meant to reduce the burden and spread of AMR. It is obvious that in order to lower infectious disease-related mortality and morbidity, an effective surveillance system for AMR, which is a component of IDSR implementation and health systems improvement, is required [7].

In a national bacteriology reference laboratory, WHO handbook seeks to give background knowledge and outline the essential stages for countries conducting AMR surveillance for meningitis, bacteremia, and common enteric epidemic-prone Diseases. The adaptation of this guide to the national context and its acceptance would help to reduce the spread of AMR among these microorganisms, even though it encourages the sharing of laboratory data on AMR for high priority bacterial infections as meningitis, cholera, salmonellosis, and shigellosis [7].

1.4 Purpose and benefits of antimicrobial stewardship programs

The goal of ASPs is aimed at improving the clinical use of prescription antibiotics, which includes proper drug choice, a sufficient/balanced range, and use at the optimal moment, in the ideal amount, using the proper technique, and at appropriate periods [5]. There is growing evidence that ASPs improve the handling of disease, reduce complications linked to the use of antimicrobial agents, and enhance patient safety and quality of care [5].

1.5 Antimicrobial stewardship programs in different parts of the world

The Infectious Diseases Study Group for Antimicrobial Stewardship (ESCMID/ESGAP) and the Antimicrobial Stewardship Working Group of the International Society of Antimicrobial Chemotherapy (ISC) collaborated to conduct a global survey on antimicrobial stewardship in hospitals between March and September 2012. There were 67 participating nations from six different continents. 103 replies out of 660 came from Central and South America. Hospital AMS requirements differed from nation to nation. A total of 46% of hospitals already have an ASP, compared to 58% of hospitals worldwide, 66% of hospitals in Europe, and 67% of hospitals in North America [5].

At the local level, some initiatives have been created, such as mandated prescriptions for the purchase of antibiotics at pharmacies. Brazil and Mexico enacted laws in 2010, Chile in 1999, Colombia in 2005, and Chile in 1999, all of which were successful in lowering antibiotic consumption. A 12% decline in penicillin consumption was realized in Mexico. Following the implementation of the legislation, the large seasonal variation in penicillin consumption also decreased, indicating that antibiotics were previously not used appropriately for the treatment of transmissible infections of the respiratory system, a common example of antimicrobials abuse. Antimicrobial resistance reduction or stabilization (87%), a decrease in the amount of antibiotic prescriptions (53%), and clinical outcome improvement (49%) were the three main goals of ASPs across all nations [5].

1.6 Antimicrobial stewardship programs and WHO Global Action Plan

The WHO Global Action Plan on Antimicrobial Resistance and the PAHO Regional Action Plan both include antimicrobial stewardship programs [5]. In fact, suggestions for maximizing the use of antimicrobial medications in human and animal health are included in Objective 4 of the WHO Global Action Plan. The same goal specifically mentions that Member States must set up stewardship programs "that monitor and promote optimization of antimicrobial use at national and local levels in accordance with international standards to ensure the correct choice of antimicrobials at the right dose, based on evidence" [5].

2. Laboratory data management system

Systems for managing and evaluating large volumes of data from antimicrobial resistance (AMR) monitoring systems are called laboratory information management systems (LIMS). A crucial tool for informing policies and infection prevention and control measures is surveillance. A Laboratory Information Management System (LIMS) is used by the majority of labs. Laboratory information management systems (LIMS) are used in real time to monitor sample flow, automatically enter laboratory results, manage quality (for as by keeping an eye on reporting times), or highlight unexpected results. To investigate AMR patterns and trends, information should even be questioned at the regional, national, and international levels. To ensure that laboratory results are directly linked to other patient data, such as admission date and outcome data, laboratory information management systems (LIMS) should ideally be able to exchange data with hospital and patient data systems [8]. The majority of labs utilize a single LIMS for all types of samples. A single sample may yield multiple results in bacteriology (positive or negative, identification of the organism or organisms, multiple AST results, and additional testing like minimum inhibitory concentration (MIC)), and the workflows are more complicated (one sample may need to be handled over the course of several days). These are frequently not the best for bacteriology because they were created for hematology and biochemistry. The actual data from potential testing, such as zone widths or MIC values, are frequently not retained by many LIMS; rather, they can only retain the interpretations, such as resilient, intermediate, or prone. It is challenging to analyze data over time and look for trends because break-points and recommendations are frequently modified [9].

2.1 Laboratory capacity and quality management

Quite frequently, the limitation in AMR surveillance is laboratory full capacity. To build this capacity and ensure quality, it is crucial to implement internal quality assurance, take part in external quality assessment (EQA) programs in research facilities, develop and keep up with national standards of operation (SOPs) to guarantee consistency and integration of lab practices, and regularly train and inspire personnel [6]. In order to perform and/or regulate the aforementioned tasks, countries should establish a central coordinating laboratory. Assuring uniformity in picking out of samples, living things, and drugs to be tested is another benefit of using monitoring criteria [6].

Working with sponsored locations to make sure they have a functional LIMS is a fundamental responsibility of the Fleming Fund Country Grants. Country Grantees must also make sure that procedures are in place that can collect data and submit it to Global Antimicrobial Resistance and Use Surveillance System (GLASS) and national surveillance systems [6].

Antimicrobial resistance (AMR) still exists in Pakistan despite the country's creation of a national action plan (NAP) to address inappropriate antimicrobial use and high antimicrobial resistance (AMR) rates there. The NAP was made public from the Department of National Health Services in May 2017. As a result, patient and LIMS are very necessary that will shape and strengthen the collection, management, and sharing of biological science information in lower middle income countries (LMICs) and boost screening of illnesses with public health significance such as antibiotic resistance (AMR). This can change how information is used to direct patient care, support information accessibility and usage for decision-making, and enable trend comparison on a regional and international level [6].

2.2 Elements of a laboratory-based surveillance system for AMR

The main method of identifying new drug resistance in the populace is surveillance, which enables prompt and effective response. Therefore, nations should improve their capability for the early detection and identification of resistant organisms that cause fetal diseases for public health [7].

Drug resistance monitoring national laboratories must staff qualified technologists, scientists, and pathologists. They should have the necessary tools and supplies to generate accurate results that will enable medication resistance surveillance. For informed decision-making, the information produced should be routinely shared with stakeholders and national authorities. Additionally, a laboratory must continually invest in purchasing supplies, media, and reagents, ensuring quality control, along with providing regular training for staff members and carrying out external quality assessment or proficiency testing, in order to successfully carry out the duties of isolation, identification, and antimicrobial susceptibility testing. A summary of the key findings from the national surveillance as well as clinical information from sentinel locations should be included in the AMR surveillance report [7]. Without requesting further information from hospitals, a nation with limited resources may carry out laboratory-based monitoring first, collecting data on susceptibility for widespread or epidemic-prone bacterial infections. Additional data may be prospectively collected if appropriate resources are available and the laboratory has experience tracking medication resistance, which should aid in enhancing AMR documentation [7].

2.3 Laboratory-based surveillance has several requirements

a. Prioritization of organisms to be observed, taking into mind the severity of the disease in the nation;

b. The selection of antibiotics to be tested for each isolate, taking into account the list of necessary medications and treatment recommendations;

c. Creating or modifying standard operating procedures (SOPs) for isolating, identifying, and testing the selected microorganisms' susceptibility to various antibiotics;

d. Implementing or improving laboratory quality systems;

e. Establishing a database for information gathering and sharing with stakeholders via already-existing systems like IDSR.

2.4 Core components of LIMS

The specimen evaluation, custom executions, and storage space group make up the three main parts of an ideal LIMS. Think of a lab where different researchers track measurements in a variety of ways, from pen and paper to a sizable spreadsheet. It would be very challenging to guarantee that human error has not tainted your data; missing data, mistakes, and discrepancies in the data collected can all exacerbate any error. Let us examine the three parts of a LIMS in greater detail and discuss how they complement one another to help study participants and lab executives (**Figure 1**) [10].

2.4.1 Recording samples using a LIMS software

The primary function of a LIMS is to record an item from the point that it arrives the laboratory by means of evaluation & preservation. In order to achieve this, all relevant evidence about the samples must be recorded at the time of first annexation, including the sample's ID, origin, collect date, and estimating information (such as concentration, volume, and particle quantity). As the laboratory item progresses over its process, further data is noted, and this data is likewise preserved in the LIMS. This comprises test results, produced data samples, and statistics for immediate research [10].

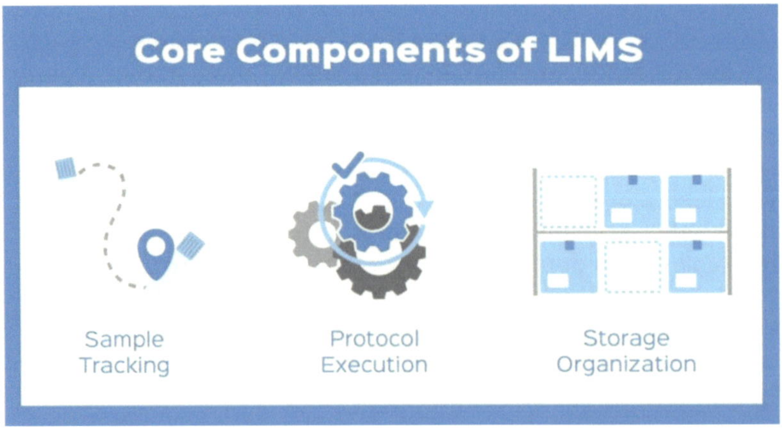

Figure 1.
Core components of LIMS.

Each lab study's specific data is recorded by a LIMS, together with details on the people involved and the environment it has been in throughout its existence. Adding a sample to a section or combining it with different materials under scrutiny in the lab are also possible. This necessitates the employment of both an exterior barcode label on the test tube and a chemical "index" in addition to it. It can therefore be differentiated from the other specimens in the pool. All of this significant sample monitoring data is kept in the LIMS technique [10].

2.4.2 Protocol execution with a LIMS

The centralization of lab's process flows and fundamental guidelines, methods, and phases is yet another key role of LIMS software. Despite of the truth, who holds the material or doing the analysis and in order to get a precise and consistent outcome, it is of the utmost importance to make sure every research individual adheres to the particulars of a given SOP (standard operating procedure). By scanning every phase in methods and standards, a LIMS supports standardization among the research group. That makes certain all of the laboratory's personnel does the right actions, in the correct order, while subject is placed accurately by means of an exam [10].

The adoption of LIMS software, which can manage trial allocations, enables an object to be assigned the appropriate procedure as soon as it reaches a testing facility. A LIMS may also provide a lab stronger code implementation restrictions, granting access to study or clinical groups, depending on who is authorized to carry out the activity. Lab test results may be taken down, transmitted through the appropriate authorization line, and then followed by relevant collaborators [10].

2.4.3 Storage organization with a LIMS

The 3rd vital part of a LIMS is to keep a record of a specimen's movements throughout its study career. The LIMS registers the location of the tube with the specimen or a container in a particular box (for example, slot A1 or B5) for each and every lab specimen. The system then keeps track of the shelf and slot that each package is on. The system also monitors the region and tier in which the chiller is located.

In busy labs, finding data fast depends on this "storage hierarchy" (Sample > Position > Box > Drawer > Rack > Shelf > Freezer > Room). Teams of researchers that are conscious of the precise location of lab samples remain useful, structured, and effective [10].

2.5 How laboratory information management system control antimicrobial resistance

LIMS can help control antimicrobial resistance in several ways:

1. Data management: LIMS can store and manage large amounts of data related to AMR, such as patient information, antibiotic usage, and susceptibility testing results. This allows for easy access, analysis, and interpretation of the data, which can help identify trends, patterns, and potential resistance patterns.

2. Antibiotic stewardship: LIMS can be integrated with antibiotic stewardship programs to help ensure proper and judicious use of antibiotics. For example,

it can provide alerts or warnings when certain drugs are being overprescribed or when there are prescribing errors. This aids in stopping the emergence and expansion of resilience.

3. Susceptibility testing: LIMS can automate and streamline the process of susceptibility testing, which determines the effectiveness of antibiotics against specific pathogens. By standardizing and automating this process, LIMS reduces the chances of human error and improves the accuracy and consistency of test results. This ensures appropriate treatment and helps prevent the misuse or overuse of antibiotics.

4. Surveillance and tracking: LIMS can facilitate the surveillance and tracking of AMR patterns within a laboratory or across multiple laboratories. It can generate reports and dashboards that provide real-time insights into resistance trends, enabling timely interventions and control measures. LIMS can also integrate with regional or national surveillance systems to enable broader data sharing and analysis for better AMR control.

5. Data sharing and collaboration: LIMS can facilitate data sharing and collaboration between different laboratories, healthcare facilities, and regulatory agencies. This allows for the exchange of information and experiences regarding resistance patterns, treatment protocols, and prevention strategies. By sharing data and collaborating, stakeholders can collectively work toward tackling AMR.

3. Conclusion

Think about combining a variety of complementary programs and technologies to increase a Laboratory Information Management Systems (LIMS) capacity for addressing issues related to antibiotic resistance and enhancing patient care. Additional software can improve data administration and analysis in addition to the drawbacks mentioned by Sayed and colleagues. These include powerful Database Management Systems (DBMS) like MySQL or PostgreSQL, machine learning and AI algorithms for predictive analytics, and data analysis and visualization tools like R or Python. While laboratory automation software improves workflow efficiency, electronic health record (EHR) integration software guarantees smooth data transmission between clinical and laboratory records. Tools for encryption and cybersecurity are necessary to protect sensitive healthcare data. While cloud computing platforms provide scalability and cost-effectiveness, geographic information system (GIS) software aids in the identification of spatial patterns. A full LIMS ecosystem includes collaboration and communication tools, regulatory compliance software, quality control and assurance tools, mobile data gathering apps, and middleware for data integration and interoperability. Solutions for data backup, user education, and documentation guarantee data integrity and user competence. The combination of these software elements enables a comprehensive strategy to tackle antibiotic resistance and improve patient care.

Conflict of interest

None of the author's interests are in conflict.

Author details

Maryam Nasrumminallah and Fatima Rehan*
Dow University of Health Sciences, Karachi, Pakistan

*Address all correspondence to: rehanfatima4@gmail.com

References

[1] Bilal H, Khan MN, Rehman T, Hameed MF, Yang X. Antibiotic resistance in Pakistan: A systematic review of past decade. BMC Infectious Diseases. 2021. Available from: https://bmcinfectdis.biomedcentral.com/articles/10.1186/s12879-02105906-1

[2] Hussain S, Khan RA, Ain NU, Haider H, Riaz S. Prevalence of extended-spectrum-β-lactamaseproducing Enterobacteriaceae: First systematic meta-analysis report from Pakistan. Antimicrobial Resist Infection Control. 2018;**7**(1):26. Available from: https://aricjournal.biomedcentral.com/articles/10.1186/s13756-018-0309-1

[3] Yasmin F, Akhtar N, Hameed A. In vitro synergistic effect of ciprofloxacin with aminoglycosides against multidrug resistant-Pseudomonas aeruginosa. Pakistan Journal of Pharmaceutical Sciences. 2013;**26**(5):1-3. Available from: https://pubmed.ncbi.nlm.nih.gov/24035966/

[4] Latif S, Anwar MS, Ahmad I. Bacterial pathogens responsible for blood stream infection (BSI) and pattern of drug resistance in a tertiary care hospital of Lahore. Biomédica. 2009;**25**(2):2-3. Available from: https://pdfs.semanticscholar.org/e089/45412f8d65ab59660d27f26cb9a9bb7cd9fd.pdf

[5] Da Silva JB Jr, Espinal M, Ramón-Pardo P. Antimicrobial resistance: Time for action. Revista Panamericana de Salud Pública. 2020;**44**:e131. DOI: 10.26633/RPSP.2020.131

[6] Malania L, Wagenaar I, Karatuna O, Andrasevic AT, et al. Setting up laboratory-based antimicrobial resistance surveillance in low- and middle-income countries: Lessons learned from Georgia. Clinical Microbiology and Infection. 2021;**27**:2

[7] Global Antimicrobial Resistance and Use Surveillance System (GLASS). Available from: https://www.who.int/initiatives/glass

[8] Tools of the Trade: Data Management Software in AMR Surveillance. 2020. Available from: https://www.flemingfund.org/publications/tools-of-the-trade-datamanagement-software-in-amr-surveillance/

[9] Fleming Fund: Supporting Surveillance Capacity for Antimicrobial Resistance. Overview of the Antimicrobial Resistance Surveillance Systems for Ghana, Malawi, Nepal and Nigeria. Available from: www.flemingfund.org

[10] Third Wave Analytics. What is a Lims and What are They Used for? Third Wave Analytics. 2023. Available from: https://thirdwaveanalytics.com/blog/what-does-a-lims-do/ [Accessed: 15 October 2023]

Section 2

Antimicrobial Resistance and Antimicrobial Stewardship

Chapter 8

Antimicrobial Resistance: A One Health Perspective in India

Radhakrishnan Rahul, Narayanasamy Damodharan, Kakithakara Vajravelu Leela, Maheswary Datchanamoorthy and Anusha Gopinathan

Abstract

One health is a collaborative, multi-sectoral, trans-disciplinary approach used to achieve optimal health and well-being outcomes that recognize the interconnections among people, animals, plants, and their shared environment. This approach is crucial because animals and people are colonized by the same bacteria species and treated with the same antibiotic classes; the technique is instrumental in fighting antibiotic resistance. The microorganism developed antibiotic-resistant genes, which were transferred to the animal and human population via the environment. Human activities speed up the organism to acquire resistance rapidly. The primary sources of antimicrobial resistance from the environment were improper sewage and hospital waste sanitation, effluents from antibiotic production units, animal husbandry waste, agricultural manure use, livestock, and aquatic sources. This study analyzed the various routes by which antimicrobial-resistant gene is transferred into humans and their pathway in India. The study concludes that implementing strict regulation and monitoring regarding the irrational use of antibiotics in animals, sewage disposal, waste disposal, and hospital infection control practices, and providing awareness to the public regarding antibiotic resistance can reduce the rate of developing antibiotic resistance to some extent along with implementing antibiotic stewardship programmes for veterinary medicine.

Keywords: antimicrobial resistance (AMR), antimicrobial stewardship (AMS), one health, environment, sources

1. Introduction

One health is a collaborative, multi-sectoral, trans-disciplinary approach working at local, regional, national, and global levels to achieve optimal health and well-being outcomes that recognize the interconnections among people, animals, plants, and their shared environment [1]. This approach is crucial because animals and people are colonized by the same bacteria species and treated with the same antibiotic classes; the technique is instrumental in fighting antibiotic resistance [2].

Antimicrobial resistance is defined as a microorganism's acquired or inherited capability to resist the effect of a particular antimicrobial and then continue its proliferation

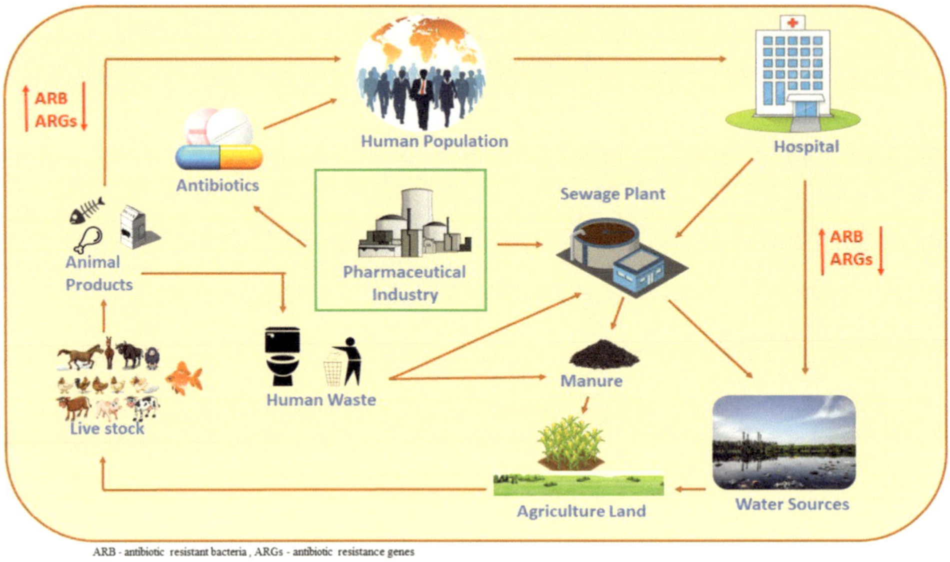

Figure 1.
Schematic diagram of antibiotic-resistant gene transfer from various sources to human.

in the host. Superbugs are microorganisms that are capable of withstanding more than one antimicrobial. At the initial stage of antimicrobial discovery, primary sources were natural sources; therefore, the chances of developing antimicrobial resistance from natural sources are now at their peak. Every organism battles with antimicrobials for survival and existence, but human actions enhance this for an organism to acquire resistance. This has led to a global threat of antimicrobial resistance and treatment failures.

As per the World Health Organization (WHO), 07 lakh deaths are reported every year due to antimicrobial resistance resulting in treatment failure. Estimates show that 10 million deaths will be reported by 2050 due to antimicrobial-resistant infections. WHO further recognizes that antibiotic resistance bacteria (ARB) and resistant genes are the major contributors to environmental pollution, and they will be a significant threat to human and animal populations worldwide.

The prevalence and occurrence of antimicrobial resistance from environmental and natural sources is not studied extensively. The environmental factors also vary across locations based on the ethnicity, culture, belief, and practices of humans worldwide. As per the existing evidence, the primary sources of antimicrobial resistance from the environment were improper sewage and hospital waste disposal, effluents from antibiotic production units, animal husbandry waste, agricultural manure use, livestock, and aquatic sources (**Figure 1**). Due to globalization, these antibiotic-resistance genes spread rapidly from one region to another. Human migration and travel, importing and exporting goods primarily carry these resistant genes and organisms to different areas [3].

2. Antibiotics in humans

Various reports across the globe state that Asian countries are the most widespread consumers of antibiotics worldwide. More specifically, India (10.7 units per person) stands first, followed by China (7.5 units per person) in antibiotic consumption. As per the latest study, the BRICS countries (Brazil, Russia, India, China, and South Africa) observed a 76% increase in the consumption of antibiotics [4]. As per the antimicrobial resistance (AMR) surveillance data from India, the most common pathogen isolated was *Escherichia coli* (*E. coli*), followed by *Klebsiella species* (22%), *Staphylococcus aureus* (*S. aureus*) (18%), *Pseudomonas species*. (10%), *Enterococcus species* (9%) and *Acinetobacter species* (8%), *Salmonella Typhi* and *Para typhi* (0.5%) [5]. The frequently consumed antibiotics in India were cephalosporins (32%), penicillin (28%) and macrolides (14%), followed by tetracyclines (6%) [6]. As per the study by ICMR, more resistance was found in cephalosporins (60%) and fluoroquinolones (41.5%) [7].

2.1 Antimicrobial resistance and human gut microbiome

The bacteria, their genomes and the environmental circumstances of the human digestive tract are referred to as the 'human gut microbiome'. The use of high-throughput, low-cost sequencing tools has fuelled research into the gut microbiome in the recent decade to determine its composition, function, and role in health and illness [8]. The gut includes hundreds of bacterial species, collectively known as the microbiota, where Bacteroidetes and Firmicutes bacteria account for 90% of all species in healthy individuals' guts [9]. Less common bacteria such as phyla Actinobacteria, Proteobacteria and Fusobacteria can protect against diseases [10, 11]. The human host has a symbiotic or commensal connection with most intestinal microorganisms. However, some organisms belonging to Enterobacteriaceae, including *Escherichia coli* and *Klebsiella pneumonia*, and *Enterococcaceae*, particularly *Enterococcus faecalis* and *Enterococcus faecium*, are found in the gut microbiota. Intestinal carriage of these bacteria can predispose to urinary tract infections and more serious systemic infections in immunocompromised people [12–14]. Infections caused by antibiotic-resistant clones of *E. coli, Klebsiella pneumoniae* [15] and *E. faecium* [16] have increased globally in recent decades. Various studies have observed that multi-drug resistance diseases cause more morbidity and mortality in low- and middle-income countries in Asia, Africa and South America [17–19].

The essential methods by which bacteria might develop resistance to antibiotics are the prevention of antibiotics reaching toxic levels inside the cell, altering the antibiotic target, and modifying or degradation of the antibiotic itself [20]. These resistance mechanisms can occur due to chromosomal gene changes and the horizontal gene transfer (HGT) of antibiotic resistance genes (ARGs) from other strains of the same or different species. HGT aided significantly in the global distribution of ARGs. HGT can occur in any environment, especially when bacterial loads are high, such as in soil [21], wastewater treatment facilities [22, 23], and human and animal gut microbiomes [24–26]. Previous research has demonstrated that the gut has many antibiotic resistance determinants, known as 'gut resistome' [27, 28].

Oral ingestion of contaminated substances (food, trash, residual waste from occupational exposure or polluted surroundings) and inhalation of airborne ARB are two routes by which ARB and ARGs reach human gut ecosystems [29, 30]. The fate of ingested bacteria, including ARB, is determined by various factors such as pre-existing microbiome structure, medication, host age and dietary environment. Under normal nutritional and physiological conditions, the diversity and abundance of the gut microbiome operate as a significant barrier to the integration of ingested bacteria, and most ingested bacteria only colonize the human body transiently. Antibiotic treatment or intake of antibiotic residues, on the other hand, may significantly modify gut microbiota structure, facilitate ARB integration by reducing competition, and enhance ARG proliferation by removing competition.

Due to a favorable environment (e.g. high concentrations of ARG, stable temperature, physiochemical conditions, and nutrient availability), long-term colonization by ARB may also accelerate HGT in the human gut and contribute to the emergence of multi-drug resistance genes by pathogens such as methicillin-resistant *Staphylococcus aureus* (MRSA).

2.2 Mechanism of horizontal gene transfer

The transfer of genetic information between bacterial species is a dynamic and continuous process resulting in the constant evolution of the bacteria. It is known to impact the host significantly with delayed or immediate effects. Most frequently recognized mechanisms of HGT occur by transduction, transformation or conjugation. Naked DNA from the extracellular environment is taken up by bacteria and integrated into their genomes during transformation. Bacteria that are naturally transformable or competent are required for the transformation process. The genes involved in DNA uptake in several other species show this trait is more prevalent [31]. Over 80 bacterial species have been experimentally demonstrated to be naturally competent. The factors that cause naturally transformable bacteria to enter a capable condition are unknown; however, nutritional deprivation and the presence of competence-inducing peptides have been identified as triggers [32]. Due to this process, competent natural pathogens such as *Neisseria gonorrhoeae*, *Vibrio cholera* and *Streptococcus pneumoniae* have acquired antibiotic resistance [33].

Membrane vesicles (MVs) are 20–250 nm spherical objects produced mainly by Gram-negative bacteria when the outer membrane bulges away from the cell and is released by constriction [34]. The cargo is delivered when MVs fuse with their target cells. MVs are generated *in vitro* by commensal bacteria in the gut. MVs formed by gut bacteria can also include cytoplasmic components, including DNA, which are significant in the context of HGT [35]. Protrusion of the outer and inner membranes is thought to create DNA-containing MVs, which leads to the incorporation of cytoplasmic components into the vesicles. Similarly, vesicle-mediated transfer of DNA has also been reported for *E. coli* [36]. While MVs are produced in the gut and can potentially influence host immune responses [37], it is unclear whether they can contribute to HGT in the gut microbiome.

Transduction is the transfer of chromosomal and extrachromosomal DNA between bacteria *via* a viral intermediary known as a bacteriophage. Generalized, specialized, and lateral transductions are the three basic transduction methods. Antibiotic-resistant genes can mobilize any portion of a bacterial genome if favorable factors exist. When bacteriophages in the lytic cycle incorporate parts of the bacterial host's DNA during capsid production, this is known as generalized transduction. Regions immediately flanking a lysogenic phage's integration site are excised and packed into the capsid in specialized transduction [38]. Prophages initiate DNA replication while still integrated into the host, known as lateral transduction.

Before excision from the host genome, this mechanism makes numerous copies of DNA. After being excised, the DNA—which can be hundreds of kilobases long and comprise both phage and neighboring genes—is packaged into new phage particles and distributed to other bacterial strains [39]. The human gut has a diverse community of bacteriophages [40, 41], and ARG-carrying phages can be found in large numbers in the gut and other settings [42]. Antibiotic therapy enhances the number of these ARG-carrying phages in the human stomach [43]. The quantitative contribution of phages to the horizontal transfer of ARGs is still unknown.

Mobile genetic elements like plasmids and integrative and conjugative elements (ICEs) are transmitted from one bacteria to another in the conjugation process [44]. Due to their large size and the common occurrence of one or more toxin-antitoxin modules that ensure that plasmids are kept inside their microbial hosts, conjugative plasmids are likely the most relevant for transmitting ARGs among conjugative elements [45].

A pilus carries DNA between bacteria close to one other during conjugation, a complicated, multi-stage, and contact-dependent process [46]. With its rich mucous coating and high density of bacterial cells, the gut provides an ideal habitat for conjugation. While colonizing the human gut, commensals and opportunistic infections have been found to transfer antibiotic resistance plasmids and ICEs [47, 48]. Notably, conjugative plasmids can supply the machinery that allows non-self-transmissible DNA to be mobilized, considerably enhancing the potential for resistance determinant HGT [49].

3. Antibiotics use in animal husbandry

In 1940, studies observed that the addition of antibiotics in cattle feed led to an enhancement in the weight gain and feed conversation ratio, which led to the irrational use of antibiotics in animals to enhance cattle growth. Over several years, this irrational use of antibiotics in cattle feed led to an increased prevalence of infectious diseases in animals due to the development of antibiotic resistance [50].

The cross-sectional study conducted in Haryana, with high milk consumption in the Indian subcontinent, showed the irrational use of antibiotics as growth enhancers and animal feed supplements. About 81% of the total participants in the study agreed that they were about to add antibiotics to improve the production of animal products. The survey indicates that 46% of the participants accepted that animal products transmitted resistant bacteria to humans. In animal husbandry, the diagnosis is critical for prescribing antibiotics [51].

The prevalence of multi-drug resistant, extended-spectrum beta-lactamases producing *Escherichia coli* from pig farms in Mizoram was studied by analyzing the fecal samples collected from organized and unorganized farms. From the analysis, they inferred that the majority of the *E. coli* were resistant to amoxicillin (81.7%), cefalexin (85.42%), co-trimoxazole (50.78%), sulfafurazole (69.38%), tetracycline (65.89%) and trimethoprim (TR) (51.94%). The results indicated the misuse and overuse of antibiotics in pig farms in India [52].

There are many possible routes for transmitting antibiotic-resistant bacteria from farm animals and their products, directly or indirectly, of which horizontal gene transfer plays a significant role [53, 54]. **Table 1** depicts various antibiotic-resistant gene observed in livestock. Shotgun metagenomics was used to analyze the characterization of the resistome, showing the presence of plasmid metagenomes in AMR genes. The most common resistance in cattle and animal products was to tetracycline, quinolone, and beta-lactams.

The behavior and destiny of ARB and ARGs discharged from animal husbandry to soil (e.g. by land application of manure and wastewater irrigation) and aquatic ecosystems have been studied extensively (e.g. through wastewater discharge and runoff).

Class	Livestock animals	Antibiotic-resistant gene identified
Penicillin	Cattle, pigs, sheep, turkeys, horses. Dogs, cats, calves.	*ampC, blaTEM, qnrS, tetW.*
Sulfonamide	Rabbits, dogs, poultry dogs, turkeys, cats.	*sulI, sulII.*
Polypeptides	Food-producing animals. Beef cattle, dairy cattle, poultry, and swine, turkey.	*mprF, yycG, rpoB, rpoC, cls2, pgsA, agrA, prs, pnpA, pmrHFIJKLM, pagP, phoP.*
Aminoglycosides	Cattle and sheep, chickens, goats, lambs, piglets, horses, turkey.	*rmtA, rmtB, armA, gar.*
Amphenicols	Poultry, birds.	*fexA, cat, cmlA, floR, cfr, fex*
Tetracycline	Calves, lambs, poultry, and swine.	*tet*
Cephalosporins	Chicks, turkey, cattle, goats, pigs, sheep.	*blaTEM-1, blaDHA-1, and cfxA.*
Polymyxins	Food-producing animals. Beef, cattle, dairy cattle, poultry, and swine.	*arnBCADTEFoperon, pmrE, pmrC, pmrA/pmrB, phoP/phoQ, mgrB, ptB, phoP/phoQ and mgrR, lpxM, acrAB, kpnEF, lpxA, lpxC, lpxD, colR/colS, cprR/cprS, parR/parS, oprH.*
Macrolides	Poultry, broilers, cattle, pigs, lambs.	*erm, carA, ole, smrB, tlrC, vgaA, vgaB, lmrA, mefA, msr, lsaA, lsaB, ereA, ereB, vgbA, vgbB, inuA, inuB, vat, mph.*
Streptogramins	Cattle, pigs swine, turkey, and broiler chickens.	*erm, carA, ole, smrB, tlrC, vgaA, vgaB, lmrA, mefA, msr, lsaA, lsaB, ereA, ereB, vgbA, vgbB, inuA, inuB, vat, mph.*
Glycopeptides	Poultry, broilers, turkeys, birds.	*Van A*
Lincosamides	Swine	*erm, carA, ole, smrB, tlrC, vgaA, vgaB, lmrA, mefA, msr, lsaA, lsaB, ereA, ereB, vgbA, vgbB, inuA, inuB, vat, mph.*
Fluoroquinolones	Dogs and cat Poultry.	*qnr*

Table 1.
Antibiotic-resistant gene observed in livestock.

Horizontal gene transfer (HGT) can transmit intracellular and free ARGs in surface and groundwater, soil and air [55–58] to indigenous bacteria [59, 60] through several paths. These ARBs may finally reach and colonize humans [21, 61], producing acute infections or long-term quiet colonization that can eventually evolve into an infection.

Bacteria-carrying ARGs are discharged into diverse receiving habitats from livestock farms *via* drainage, treated wastewater and solid waste. ARB and indigenous bacteria may undergo HGT. Humans can potentially become infected with ARB if exposed to it (e.g. water ingestion, food ingestion and inhalation). ARB might then replicate in the human body (particularly the gut) and cause endogenous or external infections.

3.1 Antibiotic use in milk animals

Raw milk is a product that is sold unprocessed. Therefore, the presence or grade of heat-treatment stages is left to the consumer's discretion. Furthermore, the consumption of non-heat-treated raw milk is now widely accepted as a trend in industrialized countries owing to its positive health effects.

The most identified antibiotics in milk animals, as per the National Dairy Research Institute (NDRI), were tetracycline, oxytetracycline, gentamicin, ampicillin, amoxicillin, cloxacillin, and penicillin due to their lower costs [62]. Studies show that farmers obtained antibiotics through various sources of consultancy. Only 50% of the small farmers consulted veterinary care for antibiotics, whereas 76.79% of medium and 87.50% of large farmers consulted the same. Other parakeet milk vendors have opted for over-the-counter medications [63]. A study conducted in 2021 at small-scale dairy farms in Assam and Haryana indicates that 53% of the small-scale farmers could not explain the term antibiotic, which stands out as an essential factor for the misuse of antibiotics in dairy farms. Residues of novobiocin, macrolides, and sulphonamides are found in the milk sample collected from the farmers [64].

The use of antibiotics and antibiotic residues in milk dates back to the 60s, indicating a swelling trend of antibiotic use in animal farms and animal products. The use of antibiotics in the milk industry may lead to antibiotic resistance and allergic reactions such as anaphylaxis and serum sickness. The antibiotic residues interact with DNA and RNA. Antibiotic residues may alter the chromosome, causing infertility due to mutation [65]. A study conducted by Modi et al. [66] studied the prevalence of *campylobacter* species in milk and milk products showing resistance against nalidixic acid, ciprofloxacin, and tetracycline. A study conducted in 2020 involved the detection of antibiotic-resistance genes in raw milk for human consumption. The traces of antibiotics in the milk clearly indicate giving antibiotics to the cattle, but there is no strict regulation to monitor and control the irrational use.

3.2 Antibiotic use in poultry

As per the reports, India was the fifth high consumer of antibiotics in animals as of 2010. However, this ratio will be doubled by 2030 [67]. Tetracycline, doxycycline, and ciprofloxacin were reported to be highly used in poultry and colistin as a growth promoter [68]. Several studies have shown traces of antibiotic residues in chicken meat intended for human consumption [69]. A survey conducted by Kaushik *et al.* in eastern India found that *E. coli* isolates showed high resistance to cefuroxime and penicillin in poultry [70].

Similarly, another study conducted in Mumbai reports a massive rise in tetracycline resistance among *Salmonella* species in poultry [71]. A recent study conducted in North India observed a high level of resistance genes against fluoroquinolones followed by tetracycline and beta-lactams [72]. Recent studies conducted in Kerala found that antibiotics were supplied along with feed, leading to antibiotic resistance [73]. Similarly, the study found *E. coli* resistance to beta-lactams and carbapenem in the poultry litter, and similar resistance has been observed and correlated with isolation from patients with UTI in that area.

Koirala *et al.* [74] conducted a cross-sectional study in 30 larger broiler poultry farms to understand the prevalence of antibiotic use and antibiotic resistance developed in poultry farms. The study found that 90% of the farms used antibiotics. Antibiotics were used for prophylaxis and therapeutic purposes. The standard categories of antibiotics used for prophylaxis are fluoroquinolone and aminoglycosides. For treatment, macrolides and polymyxin are used. In observation: 47% of them used tylosin and colistin. Neomycin and doxycycline were used in combination with antibiotic therapy. Gene mutation and enzyme production induced by chromosomes

Species	Poultry	Disease caused	Antibiotic resistance	Antibiotic-resistant gene
Staphylococcus species	Chicken, turkey.	Staphylococcus Pododermatitis Septicaemia	Beta-lactam Tetracycline Oxacillin	Altered DNA gyrase and reduced outer membrane *Mec A* resistant gene (MRSA).
Pseudomonas species	Chicken, turkey, ducks, geese, ostriches.	Sinusitis Keratitis Septicemia Endocarditis via wounds and needles for vaccines	Cephalosporins Penicillins Quinolones Monobactams	Class 1 intern to carry multiple drug-resistant genes.
Escherichia species	Chicken	Gastrointestinal illness	Tetracycline Penicillin Sulfonamides	*bla-TEM, sul2, sul3, aadA, strA, strB, tetA, tetB qnrS* resistance gene involved in transfer from farms to human population.
Salmonella species	Chicken, humans.	Gastrointestinal disease		Multi-drug resistant gene.
Streptococcus species	Cattle, pigeons, humans.	Mastitis-cattle Septicaemia-pigeons Meningitis-humans	Penicillin Tetracycline	Tetracycline resistant gene such as *tet(M), tet(L), tet(O).*
Yersinia species	Poultry meat	Acute enteritis in human *via* poultry products.	Cephalothin Ampicillin	*ystB* virulence gene.

Table 2.
Antibiotic-resistant gene observed in poultry.

within the bacteria chromosome exhibits intrinsic resistance. HGT is the reason for extrinsic or acquired resistance.

The use of antibiotics in the feed, exposure of the infected litter to the soil as manure, and consumption of meat by human can lead to the entry of resistant organisms into the human gut microbiota. **Table 2** depicts various antibiotic-resistant gene observed in poultry.

4. Pollution from the antibiotic manufacturing industry

India is the third largest country in pharmaceutical production in terms of volume, and India has 3000 manufacturing companies and 10,500 manufacturing units over the country. As per the Indian economic survey 2021, the domestic market will triple in the upcoming decade [75]. Regarding antibiotics, 80% were produced and exported globally by giant pharmaceutical manufacturers from India and China.

The studies have observed that industrial waste from these antibiotic production units has significant traces of antibiotic residues [76–78]. The flow of this antibiotic residue to the environment will be from the bulk Active Pharmaceutical Ingredients (API) manufacturing unit and the sub-manufacturing unit. The habitats impacted by pharmaceutical industrial pollution were rich in ARGs and had the highest relative abundance of ARGs of all the ecosystems studied. The relative abundances of ARGs in wastewater/sludge were also compared to the relative abundances of ARGs in most other environmental habitats (sediment, water, soil and mine), all of which are presumably less impacted by human fecal waste.

The existing Good Manufacturing Practices regulation in India mainly focuses on drug safety, raising concerns over the environmental impact. The Central Pollution Control Board (CPCB) and its constituent units scrutinize industrial waste disposal, where strict control over antibiotic residuals should also be exerted [79]. The runaway of antibiotic residue from the manufacturing companies will contaminate the soil and water sources, further enhancing antibiotic resistance development through the food chain.

5. Environmental sanitation

Better sanitation facility is a basic need of human life. Over 2.0 billion people still need to catch up on common and basic sanitation facilities such as toilets or latrines. Of these, 673 million still defecate in the open, for example, in street gutters, behind bushes or into open bodies of water [80].

Even though India has improved drastically in providing better sanitation facilities, remote rural areas still need to be adequately equipped with sanitation facilities. The association between sanitation and the occurrence of antibiotic resistance is based on the two factors, such as the antibiotic resistance prevalence rate in the area and the concentration of resistant genes in the human gut.

Although on a lower scale, ARGs can also spread through receiving water and air habitats. ARG abundances in receiving surface water habitats are typically 100 times higher than those in upstream waters. When bacteria populate the intestinal mucosa of fishes, ARGs can be obtained by bacteria reaching receiving water bodies, where they can accumulate in sediments by sedimentation and adsorption. ARGs have been found in groundwater 250 meters downstream from treatment lagoons in swine farms [81]. ARGs are also identified in aerosols up to four orders of magnitude higher than at the source, downwind of animal husbandry operations [82].

The current practice of sewage waste treatment technologies like sludge will not extensively remove the wastewater's antibiotic residues and resistance genes. The sludge produced with antibiotic residues will freely flow to the soil and environment and, finally, can enter the food chain.

6. Hospital infection control practices

Several studies have reported the resistant gene reservoir in clinical settings [83–87] occurring due to HGT, but the rate and extent of this transfer mechanism are not elaborately studied. Hospital infection control practices are pivotal in preventing antibiotic resistance in humans. Irrational use of antibiotics and poor compliance with infection control practices among healthcare providers contribute to antibiotic resistance. The lacunae in knowledge among healthcare providers regarding the growing trend of antibiotic resistance and infection control practice guidelines of the hospitals lead to irrational or overuse of antibiotics in hospitals and more frequently among independent practitioners. These poor infection control practices contribute to the rapid transmission of superbugs like MRSA and vancomycin-resistant enterococci (VRE), and other multi-drug resistant Gram-negative bacterial infections among hospital patients [88–90].

The significant challenges in these areas were the gap between the implementation and practices of infection control guidelines. Lack of knowledge regarding hand hygiene to prevent the spread of hospital infections and lack of supervision, administration, and facilities lead to the decline in infection control practices.

7. Conclusion

There is an alarming increase in antimicrobial resistance, and there is insufficient evidence regarding the prevalence and occurrence of developing antimicrobial resistance from environmental and natural sources. The environmental factors may vary across different geographical locations based on the ethnicity, culture, belief and practices of human populations. Therefore, location-based studies should be conducted to correlate the burden of ever-increasing antibiotic resistance and its impact on therapeutic outcomes. The lack of proper implementation of laws and regulations regarding the disposal of biological waste and the use of antimicrobials in the animal and agriculture field is significant contributors to the development of antibiotic resistance.

Acknowledgements

The authors thank SRM College of Pharmacy, SRM Medical College Hospital and research centre, and SRM Institute of Science and Technology, India, for providing the necessary facility for this research.

Conflict of interest

The authors declare no conflict of interest.

Author details

Radhakrishnan Rahul[1], Narayanasamy Damodharan[2], Kakithakara Vajravelu Leela[3], Maheswary Datchanamoorthy[3]* and Anusha Gopinathan[3]

1 Department of Pharmacy Practice, SRM College of Pharmacy, SRM Institute of Science and Technology, Kattankulathur, India

2 Department of Pharmaceutics, SRM College of Pharmacy, SRM Institute of Science and Technology, Kattankulathur, India

3 Department of Microbiology, SRM Medical College Hospital and Research Centre, SRM Institute of Science and Technology, Kattankulathur, India

*Address all correspondence to: drmagidatchu@gmail.com

References

[1] One health basics [Internet]. Cdc. gov.USA. 2022. [cited 2022 April 28]. Available from: https://www.cdc.gov/onehealth/basics/index.html

[2] Smillie CS, Smith MB, Friedman J, et al. Ecology drives a global network of gene exchange connecting the human microbiome. Nature. 2011;**480**:241-244

[3] O'Neill JIM. Tackling drug-resistant infections globally: final report and recommendations the review on antimicrobial resistance. 2016. Available from: https://amr-review.org/sites/default/files/160518_Finalpaper_withcover.pdf

[4] van Boeckel TP, Gandra S, Ashok A, et al. Global antibiotic consumption 2000 to 2010: An analysis of national pharmaceutical sales data. The Lancet Infectious Diseases. 2014;**14**:742-750

[5] NCDC, Containment NP on AMR, Containment NP on AMR. National Antimicrobial Resistance Surveillance Network (NARS-Net India). Delhi: National Centre for Disease Control. Available from: https://ncdc.gov.in/WriteReadData/l892s/8243867760l599818025.pdf; 2020

[6] Kotwani A, Holloway K. Trends in antibiotic use among outpatients in New Delhi, India. BMC Infectious Diseases. 2011;**11**:99

[7] Ministry of Health & Family Welfare Government of India, New Delhi. Antimicrobial Resistance and its Containment in India. 2016. Available from: https://cdn.who.int/media/docs/default-source/searo/india/antimicrobial-resistance/amrcontainment.pdf?sfvrsn=2d7c49a2_2 [Accessed 12 June 2023]

[8] Durack J, Lynch SV. The gut microbiome: Relationships with disease and opportunities for therapy. Journal of Experimental Medicine. 2019;**216**:20-40

[9] Eckburg PB, Bik EM, Bernstein CN, et al. Diversity of the human intestinal microbial Flora. Science. 1979;**2005**(308):1635-1638

[10] Ducarmon QR, Zwittink RD, Hornung BVH, van Schaik W, Young VB, Kuijper EJ. Gut microbiota and colonization resistance against bacterial enteric infection. Microbiology and Molecular Biology Reviews. 2019;**83**(3):1-29. DOI: 10.1128/MMBR.00007-19

[11] Yatsunenko T, Rey FE, Manary MJ, et al. Human gut microbiome viewed across age and geography. Nature. 2012;**486**:222-227

[12] Gorrie CL, Mirceta M, Wick RR, et al. Gastrointestinal carriage is a major reservoir of Klebsiella pneumoniae infection in intensive care patients. Clinical Infectious Diseases. 2017;**65**:208-215

[13] Price LB, Hungate BA, Koch BJ, et al. Colonizing opportunistic pathogens (COPs): The beasts in all of us. PLoS Pathogens. 2017;**13**:e1006369

[14] Raven KE, Blane B, Leek D, Churcher C, Kokko-Gonzales P, Pugazhendhi D, et al. Methodology for whole-genome sequencing of methicillin-resistant *Staphylococcus aureus* isolates in a routine hospital microbiology laboratory. Journal of Clinical Microbiology. 2019;**57**(6):1-10. DOI: 10.1128/JCM.00180-19

[15] Dunn SJ, Connor C, McNally A. The evolution and transmission of multi-drug

resistant Escherichia coli and Klebsiella pneumoniae: The complexity of clones and plasmids. Current Opinion in Microbiology. 2019;**51**:51-56

[16] Guzman Prieto AM, van Schaik W, Rogers MRC, Coque TM, Baquero F, Corander J, et al. Global emergence and dissemination of enterococci as nosocomial pathogens: Attack of the clones? Frontiers in Microbiology. 2016;7:788. DOI: 10.3389/fmicb.2016.00788

[17] Stewardson AJ, Marimuthu K, Sengupta S, et al. Effect of carbapenem resistance on outcomes of bloodstream infection caused by Enterobacteriaceae in low-income and middle-income countries (PANORAMA): A multinational prospective cohort study. The Lancet Infectious Diseases. 2019;**19**:601-610

[18] Lim C, Takahashi E, Hongsuwan M, Wuthiekanun V, Thamlikitkul V, Hinjoy S, et al. Epidemiology and burden of multidrug-resistant bacterial infection in a developing country. eLife. 2016;**5**:1-18. DOI: 10.7554/eLife.18082

[19] Gandra S, Tseng KK, Arora A, et al. The mortality burden of multidrug-resistant pathogens in India: A retrospective, observational study. Clinical Infectious Diseases. 2019;**69**:563-570

[20] Blair JMA, Webber MA, Baylay AJ, et al. Molecular mechanisms of antibiotic resistance. Nature Reviews. Microbiology. 2015;**13**:42-51

[21] Forsberg KJ, Reyes A, Wang B, et al. The shared antibiotic Resistome of soil Bacteria and human pathogens. Science. 1979;**2012**(337):1107-1111

[22] Karkman A, Do TT, Walsh F, et al. Antibiotic-resistance genes in waste water. Trends in Microbiology. 2018;**26**:220-228

[23] Che Y, Xia Y, Liu L, et al. Mobile antibiotic resistome in wastewater treatment plants revealed by Nanopore metagenomic sequencing. Microbiome. 2019;**7**:44

[24] Thanner S, Drissner D, Walsh F. Antimicrobial Resistance in Agriculture. MBio. 2016;7:2. DOI: 10.1128/mBio.02227-15

[25] San Millan A. Evolution of plasmid-mediated antibiotic resistance in the clinical context. Trends in Microbiology. 2018;**26**:978-985

[26] Huddleston JR. Horizontal gene transfer in the human gastrointestinal tract: Potential spread of antibiotic resistance genes. Infection and Drug Resistance. 2014;7:167-176

[27] Ruppé E, Ghozlane A, Tap J, et al. Prediction of the intestinal resistome by a three-dimensional structure-based method. Nature Microbiology. 2019;**4**:112-123

[28] van Schaik W. The human gut resistome. Philosophical Transactions of the Royal Society B: Biological Sciences. 2015;**370**:20140087

[29] Pérez-Cruz C, Delgado L, López-Iglesias C, et al. Outer-inner membrane vesicles naturally secreted by gram-negative pathogenic Bacteria. PLoS One. 2015;**10**:e0116896

[30] Lester CH, Frimodt-Møller N, Sørensen TL, et al. In vivo transfer of the *vanA* resistance gene from an *Enterococcus faecium* isolate of animal origin to an *E. faecium* isolate of human origin in the intestines of human volunteers. Antimicrobial Agents and Chemotherapy. 2006;**50**:596-599

[31] Johnston C, Martin B, Fichant G, et al. Bacterial transformation: Distribution, shared mechanisms and divergent control. Nature Reviews. Microbiology. 2014;**12**:181-196

[32] Chen I, Dubnau D. DNA uptake during bacterial transformation. Nature Reviews. Microbiology. 2004;**2**:241-249

[33] Blokesch M. Natural competence for transformation. Current Biology. 2016;**26**:R1126-R1130

[34] Gill S, Catchpole R, Forterre P. Extracellular membrane vesicles in the three domains of life and beyond. FEMS Microbiology Reviews. 2019;**43**:273-303

[35] Klieve AV, Yokoyama MT, Forster RJ, et al. Naturally occurring DNA transfer system associated with membrane vesicles in cellulolytic *Ruminococcus* spp. of ruminal origin. Applied and Environmental Microbiology. 2005;**71**:4248-4253

[36] Yaron S, Kolling GL, Simon L, et al. Vesicle-mediated transfer of virulence genes from *Escherichia coli* O157:H7 to other enteric Bacteria. Applied and Environmental Microbiology. 2000;**66**:4414-4420

[37] Shen Y, Torchia MLG, Lawson GW, et al. Outer membrane vesicles of a human commensal mediate immune regulation and disease protection. Cell Host & Microbe. 2012;**12**:509-520

[38] Chiang YN, Penadés JR, Chen J. Genetic transduction by phages and chromosomal islands: The new and noncanonical. PLoS Pathogens. 2019;**15**:e1007878

[39] Chen J, Quiles-Puchalt N, Chiang YN, et al. Genome hypermobility by lateral transduction. Science. 1979;**2018**(362):207-212

[40] Aggarwala V, Liang G, Bushman FD. Viral communities of the human gut: Metagenomic analysis of composition and dynamics. Mobile DNA. 2017;**8**:12

[41] Shkoporov AN, Hill C. Bacteriophages of the human gut: The "known unknown" of the microbiome. Cell Host & Microbe. 2019;**25**:195-209

[42] Debroas D, Siguret C. Viruses as key reservoirs of antibiotic resistance genes in the environment. The ISME Journal. 2019;**13**:2856-2867

[43] Fernández-Orth D, Miró E, Brown-Jaque M, et al. Faecal phageome of healthy individuals: Presence of antibiotic resistance genes and variations caused by ciprofloxacin treatment. Journal of Antimicrobial Chemotherapy. 2019;**74**:854-864

[44] Ilangovan A, Connery S, Waksman G. Structural biology of the gram-negative bacterial conjugation systems. Trends in Microbiology. 2015;**23**:301-310

[45] Pal C, Bengtsson-Palme J, Kristiansson E, et al. Co-occurrence of resistance genes to antibiotics, biocides and metals reveals novel insights into their co-selection potential. BMC Genomics. 2015;**16**:964

[46] Cabezón E, Ripoll-Rozada J, Peña A, de la Cruz F, Arechaga I. Towards an integrated model of bacterial conjugation. FEMS Microbiology Reviews. 2015;**39**(1):81-95. DOI: 10.1111/1574-6976.12085

[47] Porse A, Gumpert H, Kubicek-Sutherland JZ, Karami N, Adlerberth I, Wold AE, et al. Genome dynamics of Escherichia coli during antibiotic treatment: Transfer, loss, and persistence of genetic elements In situ of the infant gut. Frontiers in Cellular

and Infection Microbiology. 2017;**7**:1-12. DOI: 10.3389/fcimb.2017.00126

[48] Shoemaker NB, Vlamakis H, Hayes K, et al. Evidence for extensive resistance gene transfer among *Bacteroides* spp. and among *Bacteroides* and other genera in the human Colon. Applied and Environmental Microbiology. 2001;**67**:561-568

[49] Ramsay JP, Firth N. Diverse mobilization strategies facilitate transfer of non-conjugative mobile genetic elements. Current Opinion in Microbiology. 2017;**38**:1-9

[50] von Wintersdorff CJH, Penders J, van Niekerk JM, Mills ND, Majumder S, van Alphen LB, et al. Dissemination of antimicrobial resistance in microbial ecosystems through horizontal gene transfer. Frontiers in Microbiology. 2016;**7**:173. DOI: 10.3389/fmicb.2016.00173

[51] Parkunan T, Ashutosh M, Sukumar B, et al. Antibiotic resistance: A cross-sectional study on knowledge, attitude, and practices among veterinarians of Haryana state in India. Veterinary World. 2019;**12**:258-265

[52] Mandakini R, Roychoudhury P, Subudhi PK, et al. Higher prevalence of multidrug-resistant extended-spectrum β-lactamases producing Escherichia coli in unorganized pig farms compared to organized pig farms in Mizoram, India. Veterinary World. 2020;**13**:2752-2758

[53] Economou V, Gousia P. Agriculture and food animals as a source of antimicrobial-resistant bacteria. Infection and Drug Resistance. 2015:49-61

[54] Rossi F, Rizzotti L, Felis GE, et al. Horizontal gene transfer among microorganisms in food: Current knowledge and future perspectives. Food Microbiology. 2014;**42**:232-243

[55] Han X-M, Hu H-W, Chen Q-L, et al. Antibiotic resistance genes and associated bacterial communities in agricultural soils amended with different sources of animal manures. Soil Biology & Biochemistry. 2018;**126**:91-102

[56] Koike S, Krapac IG, Oliver HD, et al. Monitoring and source tracking of tetracycline resistance genes in lagoons and groundwater adjacent to swine production facilities over a 3-year period. Applied and Environmental Microbiology. 2007;**73**:4813-4823

[57] Fang H, Han L, Zhang H, et al. Dissemination of antibiotic resistance genes and human pathogenic bacteria from a pig feedlot to the surrounding stream and agricultural soils. Journal of Hazardous Materials. 2018;**357**:53-62

[58] Sancheza HM, Echeverria C, Thulsiraj V, et al. Antibiotic resistance in airborne Bacteria near conventional and organic beef cattle farms in California, USA. Water, Air, and Soil Pollution. 2016;**227**:280

[59] Zhang Y, Snow DD, Parker D, et al. Intracellular and extracellular antimicrobial resistance genes in the sludge of livestock waste management structures. Environmental Science & Technology. 2013;**47**:10206-10213

[60] Fang H, Wang H, Cai L, et al. Prevalence of antibiotic resistance genes and bacterial pathogens in long-term manured greenhouse soils As revealed by metagenomic survey. Environmental Science & Technology. 2015;**49**:1095-1104

[61] Zeng J, Pan Y, Yang J, et al. Metagenomic insights into the distribution of antibiotic resistome between the gut-associated environments and the pristine environments. Environment International. 2019;**126**:346-354

[62] Grover CR, Bhavadasan MK. Antibiotic Residues in Milk: A Public Health Concern. Haryana: National Dairy Research Institute; 2020. Available from: https://cdn.cseindia.org/userfiles/antibiotic_milk_chand%20ram.pdf [Accessed 12 June 2023]

[63] Kumar V, Gupta J. Prevailing practices in the use of antibiotics by dairy farmers in eastern Haryana region of India. Veterinary World. 2018;**11**:274-280

[64] Kumar N, Sharma G, Leahy E, et al. Understanding antibiotic usage on small-scale dairy farms in the Indian states of Assam and Haryana using a mixed-methods approach—Outcomes and challenges. Antibiotics (Basel). 2021;**10**:1-13

[65] Sachi S, Ferdous J, Sikder MH, et al. Antibiotic residues in milk: Past, present, and future. Journal of Advanced Veterinary and Animal Research. 2019;**6**:315-332

[66] Modi S, Brahmbhatt MN, Chatur YA, et al. Prevalence of Campylobacter species in milk and milk products, their virulence gene profile and anti-bio gram. Veterinary World. 2015;**8**:1-8

[67] Tóth AG, Csabai I, Krikó E, et al. Antimicrobial resistance genes in raw milk for human consumption. Scientific Reports. 2020;**10**:7464

[68] van Boeckel TP, Brower C, Gilbert M, et al. Global trends in antimicrobial use in food animals. Proceedings of the National Academy of Sciences of the United States of America. 2015;**112**:5649-5654

[69] Brower CH, Mandal S, Hayer S, et al. The prevalence of extended-Spectrum Beta-lactamase-producing multidrug-resistant Escherichia Coli in poultry chickens and variation according to farming practices in Punjab, India. Environmental Health Perspectives. 2017;**125**:77015

[70] Gandra S, Joshi J, Trett A, Lamkang AS, Laxminarayan R. Scoping Report on Antimicrobial Resistance in India. Washington, DC: Center for Disease Dynamics, Economics & Policy; 2017. Available from: https://onehealthtrust.org/wp-content/uploads/2017/11/scoping-report-onantimicrobial-resistance-in-india.pdf [Accessed 12 June 2023]

[71] Ramakant Sahu, Poornima Saxena. Antibiotics in Chicken Meat. New Delhi: Centre for Science and Environment; 2014. Available from: https://cdn.cseindia.org/userfiles/Antibiotics%20in%20Chicken_Lab%20Report_Final%2029%20July.pdf [Accessed 12 June 2023]

[72] Kaushik P, Anjay A, Kumari S, et al. Antimicrobial resistance and molecular characterization of E. coli from poultry in eastern India. Veterinaria Italiana. 2018;**54**:197-204

[73] Sebastian S, Tom AA, Babu JA, et al. Antibiotic resistance in Escherichia coli isolates from poultry environment and UTI patients in Kerala, India: A comparison study. Comparative Immunology, Microbiology and Infectious Diseases. 2021;**75**:101614

[74] Koirala A, Bhandari P, Shewade HD, et al. Antibiotic use in broiler poultry farms in Kathmandu Valley of Nepal: Which antibiotics and why? Tropical Medicine and Infectious Disease. 2021;**6**:47

[75] Pharmaceuticals Export Promotion Council IQVIA A-A, (PIB) PIB, Pharmaceuticals Export Promotion Council IQVIA A-A. Indian Pharmaceuticals Industry Report.

New Delhi: Department for Promotion of Industry and Internal Trade (DPIIT); 2021

[76] Gothwal R, Shashidhar. Occurrence of high levels of fluoroquinolones in aquatic environment due to effluent discharges from bulk drug manufacturers. Journal of Hazardous, Toxic, and Radioactive Waste. 2017;**21**:5016003

[77] Larsson DG, de Pedro C, Paxeus N. Effluent from drug manufactures contains extremely high levels of pharmaceuticals. Journal of Hazardous Materials. 2007;**148**:751-755

[78] Lübbert C, Baars C, Dayakar A, et al. Environmental pollution with antimicrobial agents from bulk drug manufacturing industries in Hyderabad, South India, is associated with dissemination of extended-spectrum beta-lactamase and carbapenemase-producing pathogens. Infection. 2017;**45**:479-491

[79] Ministry of Environment & Forests. Bio-medical Waste Management Rules 2016. New Delhi: Government of India; 2016

[80] Kizny Gordon AE, Mathers AJ, Cheong EYL, et al. The hospital water environment as a reservoir for Carbapenem-resistant organisms causing hospital-acquired infections—A systematic review of the literature. Clinical Infectious Diseases. 2017;**64**:1435-1444

[81] Breathnach AS, Cubbon MD, Karunaharan RN, et al. Multidrug-resistant Pseudomonas aeruginosa outbreaks in two hospitals: Association with contaminated hospital wastewater systems. Journal of Hospital Infection. 2012;**82**:19-24

[82] Hota S, Hirji Z, Stockton K, et al. Outbreak of multidrug-resistant *Pseudomonas aeruginosa* colonization and infection secondary to imperfect intensive care unit room design. Infection Control and Hospital Epidemiology. 2009;**30**:25-33

[83] Paterson DL, Bonomo RA. Extended-Spectrum β-lactamases: A clinical update. Clinical Microbiology Reviews. 2005;**18**:657-686

[84] Weingarten RA, Johnson RC, Conlan S, Ramsburg AM, Dekker JP, Lau AF, et al. Genomic analysis of hospital plumbing reveals diverse reservoir of bacterial plasmids conferring Carbapenem resistance. MBio. 2018;**9**(1):1-16. DOI: 10.1128/mBio.02011-17

[85] Rowe WPM, Baker-Austin C, Verner-Jeffreys DW, et al. Overexpression of antibiotic resistance genes in hospital effluents over time. The Journal of Antimicrobial Chemotherapy. 2017;**72**:1617-1623

[86] Acharya AS, Khandekar J, Sharma A, et al. Awareness and practices of standard precautions for infection control among nurses in a tertiary care hospital. The Nursing Journal of India. 2013;**104**:275-279

[87] Goyal A, Narula H, Gupta PK, et al. Evaluation of existing knowledge, attitude, perception and compliance of hand hygiene among health care workers in a tertiary care Centre in Uttarakhand. Journal of Family Medicine and Primary Care. 2020;**9**:1620-1627

[88] Sodhi K, Shrivastava A, Arya M, et al. Knowledge of infection control practices among intensive care nurses in a tertiary care hospital. Journal of Infection and Public Health. 2013;**6**:269-275

[89] Taneja J, BibhaBati M, Aradhana B, et al. Evaluation of

knowledge and practice amongst nursing staff toward infection control measures in a tertiary care hospital in India. The Canadian Journal of Infection Control. 2009;**24**:104-107

[90] Tenna A, Stenehjem EA, Margoles L, et al. Infection control knowledge, attitudes, and practices among healthcare workers in Addis Ababa, Ethiopia. Infection Control and Hospital Epidemiology. 2013;**34**:1289-1296

Chapter 9

Antimicrobial Resistance in Environment and Antimicrobial Stewardship

Sadia Khan

Abstract

The spread of antimicrobial resistance (AMR) in the environment is an alarming issue for the world as the extensive use of antimicrobials in different sectors including healthcare facilities, food and pharmaceutical industries, agriculture, and animal farming has resulted in the enrichment of antibiotic-resistant bacteria (ARB) and antibiotic resistance genes (ARGs) in different environmental compartments such as surface water, wastewater, soil, and drinking water. Not only single-drug-resistant but multidrug-resistant (MDR) organisms are increasing at an alarming rate. Treatment technologies used in wastewater treatment plants (WWTP) are mostly focused on the removal of physical and chemical contaminants and less focused on the removal of biological contaminants like antimicrobial-resistant genes, which pose serious threats for both humans and the environment. Antimicrobial stewardship (AMS) programs have been started in different countries of the world to overcome the problem of antimicrobial resistance and minimize the impacts on the environment. This program is based on collective efforts from clinicians, technicians, physicians, scientists, leaders, and the public and their active participation in the possible eradication of antimicrobial resistance from the world.

Keywords: multidrug-resistant bacteria, wastewater treatment plant, chlorine, horizontal gene transfer, antimicrobial stewardship

1. Introduction

With the production of antibiotics in the twentieth century, the world is facing the dilemma of AMR, which decreases the efficiency of antimicrobial compounds for the treatment of infectious diseases. Antimicrobial resistance (AMR) has been recognized as a global threat and the endless emergence of multi-drug resistant strains of bacteria like New Delhi Metallo-β-lactamase-1 (NDM-1) containing *Enterobacteriaceae*, Multidrug-Resistant *Staphylococcus aureus* (MRSA), and Extensively Drug-Resistant Tuberculosis (X-DRTB) become a complicated challenge worldwide [1, 2]. The resistant organisms and their genes have been dispersed in different environments due to the widespread use of antimicrobials to avoid the onset of infections in humans and animals. The uncontrolled and unsafe use of antibiotics since the beginning resulted in the emergence of resistance in almost every environment including drinking water, wastewater, soil, and even in pristine environments. Antibiotic resistance is so established in each environment that the analysis of bacteria for antibiotic resistance genes reveals the source of bacteria due to the use of different antibiotics for humans, livestock, and agriculture [3].

Antimicrobial resistance is not only developed in organisms when they are exposed to antimicrobial agents in the natural environment but also without any contact with antibiotics in the form of intrinsic resistance. Furthermore, not only the higher concentrations of antibiotics but also minimum inhibitory concentration (MIC) could enrich the antibiotic-resistant bacteria and their genes in the environment [4]. The high level of ARB and ARGs is due to anthropogenic activities and needs immediate attention as the use of antibiotics is continuously increasing, and the development of the next genera of antibiotics is limited.

2. Wastewater and antimicrobial resistance

Wastewater is one of the major sources of ARB and ARGs in the environment. The higher concentrations of antibiotics in wastewater cause the enrichment and propagation of resistance among other bacteria and also transfer resistance to another environment [5]. The discharge of untreated wastewater has resulted in the enrichment and prevalence of ARGs in the water environment and the wastewater treatment plants (WWTP) act as a major reservoir of ARB and ARGs as activated sludge and treated effluent contain ARB and ARGs in abundance [6]. Activated sludge faces a lot of stress and induces the AMR through the co-selection process in the wastewater environment [7].

Wastewater contains a variety of pollutants and diverse microbial communities and provides a favorable environment for the proliferation of ARGs among bacteria. The major sources of antimicrobial resistance are humans and animals. Discharges from clinical and industrial sources also contribute to the ARB and ARGs. These pollutants move with the water cycle and reach other water environments [5]. Natural water bodies now have higher concentrations of antimicrobial compounds in water and their sediments. Treatment technologies used in WWTP are not focused on the removal of ARB and ARGs, and they remain active in effluent. During wastewater treatment, primary, secondary, and tertiary treatments have been used to remove most of the pollutants, and disinfection is also applied at the last stages of the treatment. Biological treatments including aerobic and anaerobic bioreactors and constructed wetlands are used to remove resistant bacteria. The concentrations of commonly used antibiotics are much higher in wastewater and surface water indicating the possible transfer of these resistances to other environments as well as due to the use of polluted water or untreated or partially treated wastewater in agriculture and for plantation in many countries.

3. Drinking water and antimicrobial resistance

Physical, chemical, and biological environmental factors affect the occurrence and enrichment of ARB and ARGs in drinking water including the disinfectants such as chlorine and pipe materials of the distribution system [8]. Wastewater is considered a hotspot for the ARB and ARGs due to the presence of human excreta and chemicals that contribute to the emergence of resistance in the bacterial population, although drinking water can also harbor the same [9] despite having different physico-chemical and biological characteristics. Drinking water supply systems contain pathogens and resistant bacteria that can transfer genes to other populations. The water distribution system is considered a complex system, and it is difficult to inactivate and treat the resistant bacteria as complete information about the behavior of these bacteria is not well known. Disinfectants used for the treatment of pathogenic microorganisms can result in the proliferation of ARB and their genes in the water distribution system and drinking water.

Chlorine is a cheap and effective disinfectant for the elimination of harmful microorganisms, especially pathogens from drinking water [10]. It is used as residual disinfection, and it reacts with the organic matter in water and produces undesirable disinfectant by-products (DBPs). Several DBPs can be found in the treated drinking water at much higher concentrations; trihalomethane (THM), haloacetic acid HAA, dichloroacetic acid (DCAA), trichloroacetic acid (TCAA), dibromochloromethane (DBCM), trichloromethane (TCM), tribromomethane (TBM), bromodichloromethane (BDCM). Their concentration in drinking water depends on the chlorine dose, temperature, and pH of the water. They can cause several health effects; adverse birth outcome, growth reduction in infants, mutagenesis, bladder and colon cancer (carcinogenic), and neurotoxicity but the extent of adverse effects depends on the concentration of the DBP and the exposure duration, which in some cases exceeded up to 40 years [11, 12].

The response of antimicrobial compounds such as chlorine and monochloramine in the main distribution system and the premise plumbing system could be different: They disappear more quickly from the premise plumbing than the main system. As a reactive chemical and strong oxidizing agent, chlorine can react with plumbing materials such as copper [13, 14]. The concentration decreases and this sub-lethal concentration can cause the selection and enrichment of ARB. The minimum selective concentration of disinfectants exerts selective pressure on the resistant population to survive and proliferate in the distribution system thus enriching the ARB and ARGs [8].

Opportunistic pathogens are an emerging water-borne issue. They are not contaminants. They are a normal habitant of drinking water, but they can cause life-threatening human diseases with an economic loss of one billion dollars annually. Their common characteristics help them to proliferate in the drinking water distribution and plumbing systems. They are oligotrophic and can grow at low organic carbon. They are persistent in drinking water, have thermal tolerance, and can grow in amoebae and stagnant water [15].

Common sources of waterborne diseases from opportunistic pathogens are *Legionella pneumonia*, *Mycobacterium avium*, and *Pseudomonas aeruginosa* [16]. Besides them, other bacteria are also involved in these outbreaks such as *Methylobacterium*, *Acinetobacter*, and *Aeromonas*. *M. avium*, an opportunistic pathogen of household plumbing and causes pulmonary infections [17], is resistant to disinfectants, high temperatures, phagocytic amoebae, and tolerates low oxygen concentration [18]. These antimicrobial-resistant bacteria are posing a threat to the public.

A consistent habitant of the drinking water plumbing system is *Legionella pneumophila*, a waterborne pathogen and causative agent of the life-threatening lung disease "Legionnaires" [19, 20]. Drinking water species are more chlorine-resistant than laboratory-grown cultures because of their presence in amoebae and biofilms on pipes. In the United States, Legionella is the most prevalent pathogen of waterborne infection outbreaks. Another organism that is associated with tap water systems and causes pneumonia is *Pseudomonas aeruginosa* [21]. They are resistant to chlorine and antibiotics and can utilize nitrates, which helps them to survive in stagnant water, form biofilms [22, 23], and grow well in flowing water in pipes [24].

Opportunistic pathogens found in premise plumbing systems share some common features and have a selective advantage over other competitors. Their resistance to disinfectants and phagocytosis, the ability of biofilm formation and regrowth, and tolerance of low oxygen concentrations allow their enrichment in drinking water distribution systems posing a health hazard for the population. Slow growth also facilitates their survival as they skip during disinfection, and their death rate is also diminished due to the growth rate. As normal inhabitants of drinking water, their concentration does not decrease when they move from the source as compared to contaminants [20, 25].

Bacteria added to the water from the environment cause water contamination. *Escherichia coli* and *Salmonella* species are well-known drinking water contaminants. There is no correlation between the presence of opportunistic pathogens and fecal indicator organisms in water. The advancement in treatment technologies resulted in a selective environment for opportunistic pathogen growth in drinking water distribution and plumbing systems, and state-of-the-art technologies are needed to overcome the problem in drinking water. The premise plumbing system is a complex and variable environment and difficult to control thus can easily cause an outbreak of a water-borne disease [20].

One recommended method to kill these pathogens is to increase the concentration of disinfectants, but it proves to be ineffective as resistance has been developed against the higher doses of the disinfectants. The same results were observed with disinfection substitution. Turbidity reduction can also reduce the number of these bacteria such as *Mycobacterium avium*. An increase in the temperature of the water could also result in the inhibition of the growth of some bacteria.

4. Agriculture, soil, and animal farming

In soil, antibiotics are absorbed on the organic residues and also undergo transformation and biodegradation thus increasing their persistence in the soil environment. Their strong binding with soil organic components resulted in stable residues that persist in the agro-environment for a longer period. They can make enzymatic changes in local microorganisms and alter their metabolic activities for different nutrient resources such as carbon. Additionally, they manipulate microbial biomass and relative abundances of different microbial species [26].

Landfilling sites that receive municipal solid waste as disposed material contain large amounts of pharmaceutical and personal care products; these are the key anthropogenic sources of antibiotics in soil [27]. The leachate from the landfilling sites serves as the hotspots for the dissemination and enrichment of ARB and ARGs. An important source of the development of ARB is the agroecosystem, which has a direct effect on human leaching from agricultural soil resulting in the contamination of surface water [28]. In agriculture, feces-contaminated manure is used, which can be a source of resistant development in soil bacteria as repeated exposure to sub-inhibitory concentrations of antibiotics provokes resistance in them [29].

Different factors might be involved in the overall occurrence of antimicrobial resistance in the agro-environment. The physicochemical properties of the antibiotic residues, soil characteristics, and climate factors such as rainfall, temperature, and humidity all could contribute to the persistence of antibiotic resistance in the soil environment [30]. Microbial colonization is very diversified in soil, which serves as a reservoir of resistance genes. A complete insight into the gene transfer mechanism is required to understand the mechanism [31]. Horizontal gene transfer (HGT) and co-selection are the two major methods of antimicrobial-resistant gene transfer in soil environments [32, 33]. Horizontal gene transfer is carried out by three methods in a bacterial cell, which are transduction, conjugation, and transformation. Transduction is the transfer of genetic material through a virus, conjugation is the direct transfer of genetic material from one bacterium to another, and in transformation bacteria directly take up foreign or exogenous genetic material from the environment without any direct contact with other bacteria (**Figure 1**).

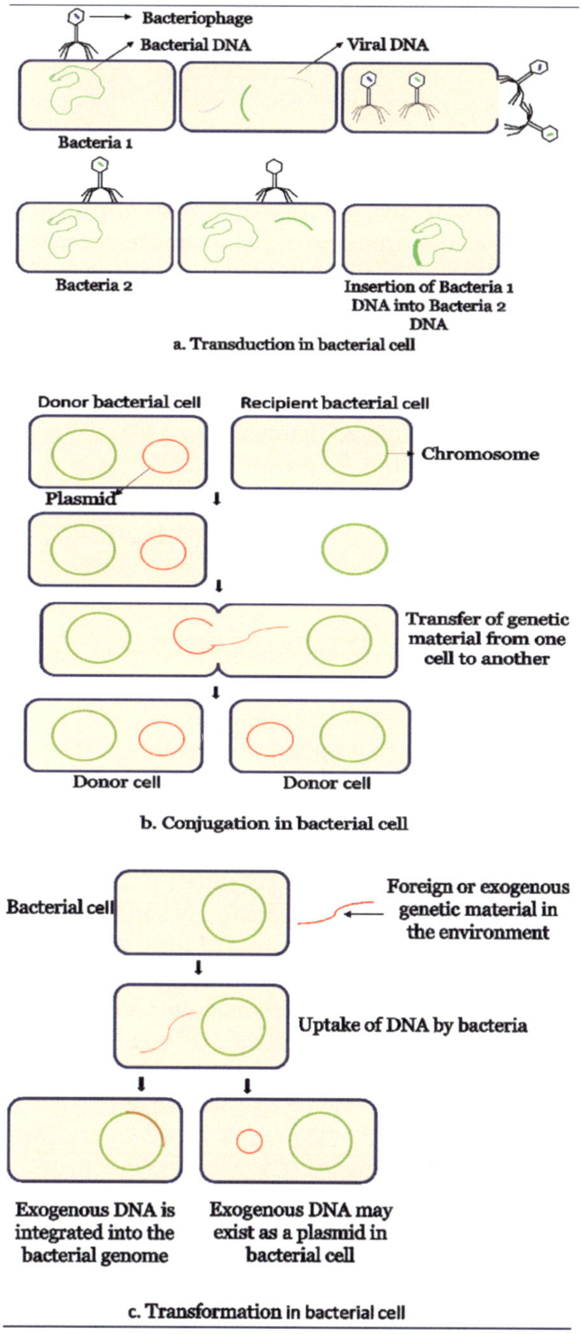

Figure 1.
Horizontal gene transfer in bacterial cells. a. transduction, b. conjugation, c. transformation.

Antibiotics have been used extensively in different industries causing environmental contamination and the development of resistant organisms with human health impacts, such as increased use of antibiotics in animal farming could result in the selection of organisms with resistance to human-used antibiotics [29]. The European Union banned the use of growth-promoting antibiotics in the veterinary industry in 2006 [34]. Antibiotic use in animal husbandry has not only a higher potential for water pollution but also more ecotoxicological risks [29].

5. Antibiotic stewardship

To combat the complicated AMR, the widely used method is antimicrobial stewardship (AMS), which emphasizes minimized and controlled use of antibiotics in different fields and prevents the exposure of pathogens in the natural environment. Antibiotic stewardship is responsible and careful management of antibiotic use [35]. It is also mandatory to avoid the over-prescription of antibiotics by healthcare professionals as they are the major driving force for the misuse of antibiotics. Antibiotic stewardship programs educate them about the overuse of antibiotics and their impact on human beings in the natural environment. The incidences of overtreatment with multiple antibiotics must be minimized in healthcare [36] as this practice evolved AMR into harmless bacteria. Antibiotic stewardship programs also focus on the reduced cost of treatment and minimize the economic impact of AMR. A situational analysis is performed before starting AMR stewardship, and it consists of strengths, weaknesses, opportunities, and threats (SWOT) of any facility [37]. It identifies the possible barriers to the implementation of the AMS program and facilitators who can contribute to the process of AMS.

Broad-spectrum antibiotics are used in healthcare for the treatment of infectious diseases but to defeat the AMR, narrow-spectrum antibiotics becomes the choice along with the use of combinatorial drug therapy as no single antibiotic is found to be effective for the treatment of all infectious diseases [38]. Combination drug therapy has more efficiency than single antibiotics against bacterial infections as they are based on synergism and antagonism [39]. Contrary to this, narrow-spectrum antimicrobials can put immense selective pressure on bacteria and can cause serious complications. Similarly, combinatorial drug therapy is expensive because two or more antibiotics are used in treatment although it is one of the most reliable and effective methods for the treatment of infectious diseases. This strategy is more effective for gram-negative bacteria than for gram-positive bacteria.

Several programs have been started by governments and agencies in developed countries for the eradication of AMR such as the National Antimicrobial Resistance Monitoring System for Enteric Bacteria (NARMS) by the Centers for Disease Control and Prevention (CDC) in the USA, which was established in 1996 and provides information about the emerging bacterial resistance. Canadian Integrated Program for Antimicrobial Resistance Surveillance (CIPARS) started in 2002 and collects information about the current trends in antimicrobial use and AMR in enteric bacteria to prolong the effectiveness of antimicrobial drugs in humans and livestock. The Danish Integrated Antimicrobial Resistance Monitoring and Research Programme (DANMAP) in Denmark, working since 1995, reviews the use of antibiotics in different sectors, the occurrence of antimicrobial resistance, the relationship between the use of antibiotics and the development of resistance, and the route of transmission to minimize the spread of AMR [1].

Although the use of some antibiotics has decreased in the world, on the other hand, the use of other antibiotics has increased. This resulted in a decrease in resistance to one class of antibiotics but an increase in resistance to another class of antibiotics. There is no alternative drug for some antibiotics as pathogens have developed resistance against many broad-spectrum antibiotics. Many recently developed antimicrobial resistance genes have been identified such as Mobile Colistin Resistant Genes (MCR) for Colistin, which are isolated from multiple environmental sources including sea, rivers, and sewage wastewater [40].

Regular monitoring of the use of antibiotics in different sectors and their discharge into the natural environment must be effectively examined to prevent the spread of AMR in these environments. A hundred percent effectiveness of any program can be achieved by proper implementation of rules, regulations, public laws, and policies, especially those related to public health and the environment. Uninterrupted data inspection is needed for the use of antibiotics, development of resistance, resistant patterns of antibiotics, resistance to other drugs, and outbreak of bacterial diseases. National and international coordination with proper guidance for health organizations is necessary for antimicrobial stewardship as it becomes a challenge for developing countries that lack infrastructure and good governance. Developed countries ensure the antimicrobial stewardship implementation program to eradicate AMR.

Another issue facing the world regarding the eradication of AMR in the environment is the lack of discovery for advanced antibiotics as a drastic decline has been observed since 2000. The development of antibiotics is a time-consuming process that involves research, clinical trials, approvals, and production costs; hence pharmaceutical industries are not taking much interest in research and development for brand-new drugs. Bacteria have shown AMR against most of the antibiotics used in the world, and highly potent antibiotics are urgently needed to reduce the outbreaks of diseases [39]. Innovation is another factor for the development of effective antibiotics as most industries are using already well-established compounds. Only one or two potentially active compounds against gram-negative bacteria are expected in the next decade, although a lot of research is going on for this, but the success rate is low. Extensive investment should be made by the industries to solve the issue. Governments must declare it as a priority and provide the funding and human resources for AMR stewardship in their countries.

Increased use of antibiotics is prevalent not only in humans but also in animals for livestock management and food production, which resulted in the fixation of AMR in bacteria. These bacteria can transfer resistant genes from animal to human by direct or indirect contact. Direct contact involves the interaction with animals and their biological fluids such as blood, urine, milk, and saliva whereas indirect contact may involve the use of contaminated food such as dairy products, eggs, and meat. Thus, zoonotic animals are also a reservoir of AMR just like humans including adults, children, and infants. An important factor that contributes to the increased use of antibiotics is the over-the-counter availability of antibiotics in many countries. Co-selection of resistant genes is another important factor contributing to the spread of AMR in bacteria as it causes resistance to many antibiotics simultaneously, which belong to different classes of antibiotics [41].

Public awareness about any problem plays a vital role in the management and control of that problem. There is a lack of public awareness among people and even among practitioners and doctors who intentionally or forcefully over-prescribe antibiotics for treatment on the demands of pharmaceutical companies. Clinicians use higher doses of antibiotics than required and also prescribe them for a longer period than necessary [2]. Optimum use of antimicrobial compounds could also help to reduce the development of AMR. It is essential to start public awareness about AMR on a wider scale if we want to prevent it further. Similarly, public participation is a key factor in such stewardship programs as the active contribution of the general population makes the program successful.

Healthcare workers and policymakers must be encouraged to adapt best practices to combat AMR. Effective communication and training must be provided to the people working in the healthcare system so that they can educate others. Extensive research is needed to understand the mechanisms of development of AMR in the natural environment. Personal sanitation and hygiene should also be encouraged for infection prevention.

All forms of life from the smallest to largest organisms have defense mechanisms in the form of antimicrobial peptides, which have broad-spectrum activity against prokaryotic and eukaryotic microorganisms [39]. Due to their characteristics, they can be a useful alternative to currently used antibiotics [42]. An emerging method to control AMR is the use of bacteriophages (BP) for phage therapy [38, 43]. It uses natural or bioengineered phages for the lysis of disease-causing bacteria (**Figure 2**). It has many advantages over other methods such as lack of cross-resistance for other antibiotics, strain-specific, and effectiveness against a wide range of bacteria. They can be used individually or in combination with other antibiotics [43]. Small particles that are favorable to use against AMR are nanoparticles (NPs) as they are target-specific and stabilize the drugs inside the body [38]. They can help in horizontal gene transfer, and their use can cause the emergence of resistance in other bacteria in the environment [44].

The bioactive phytochemicals in plants that belong to different chemical classes can also be used against AMR as they have bactericidal properties and can reverse AMR [45]. They have synergistic effects with other antibiotics and can serve as promising solutions for AMR. Biosurfactants have anti-adhesive properties. They inhibit bacterial colonization and have shown effectiveness against multidrug-resistant *Bacillus subtilis*. Quorum sensing is a method of bacterial communication and is used for bacterial colonization; by inhibiting the quorum sensing with some

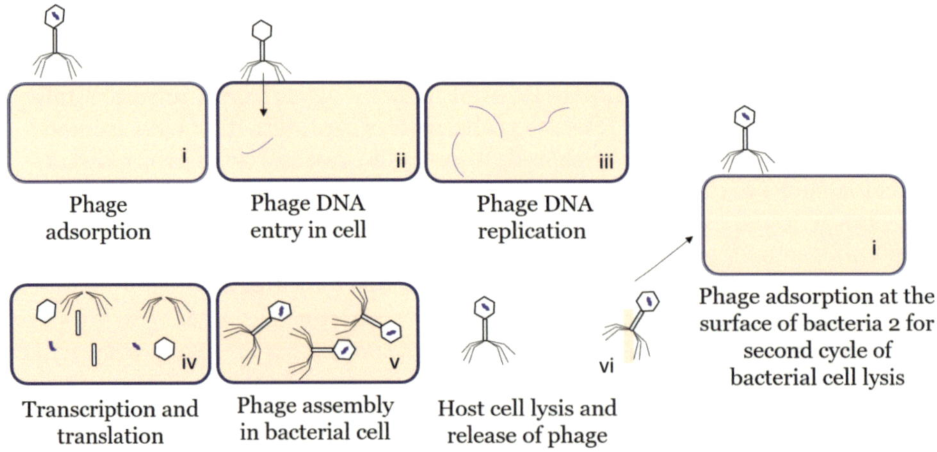

Figure 2.
Phage therapy as an alternative to antibiotics for the treatment of bacterial infections.

inhibitors bacterial colonization can be restricted [46, 47]. The process is called Quorum quenching. Quorum quenching weakens the signals for bacterial colonization in the environment and decreases the pathogenicity of the microbes with the help of Quorum sensing inhibitors (**Figure 3**). Postbiotic compounds such as proteins, carbohydrates, lipids, vitamins, and organic acids can unambiguously manipulate microbial and host functions. This method is not globally accepted and needs further research.

Antimicrobial stewardship programs are based on three pillars, which include the strengthening of the health system by an integrated approach, patient safety, and infection prevention and control. It applies to human, animal, food, agriculture, and pharmaceutical sectors. Proper technology and data must be available to cope with AMR. State-of-the-art laboratories with advanced equipment and other facilities must be present for the on-time diagnosis and treatment of infectious diseases [38]. Furthermore, patients' records must be maintained who have been treated for multidrug-resistant infections or using antibiotics for a longer period [39]. AMR eradication from the natural environment is a slow and steady process and cannot be achieved overnight.

Leadership commitment plays a vital role in the success of any program and antibiotic stewardship also needs this commitment from leadership, healthcare workers, and physicians. Production and distribution records of antibiotics must be maintained by the government and healthcare facilities. Similarly, regulations and responsibilities should be documented to avoid the overuse of antibiotics in the healthcare system [39]. There must be accountability and proper distribution of responsibilities to achieve the target. A multidisciplinary team is needed, which may consist of physicians, microbiologists, nurses, lab technicians, pharmacists, and experts in infection management [42, 48] with a fast-track approach of rapid diagnosis, testing, and reporting of results [49].

AMR is a global concern and needs novel, innovative, and permanent solutions as the burden of the AMR economy is huge and needs to be solved immediately. Both developed and underdeveloped countries must work together to overcome the obstacles. The development of potent novel antimicrobial compounds and monitoring of antibiotic abuse are mandatory to win the fight against AMR [38, 42].

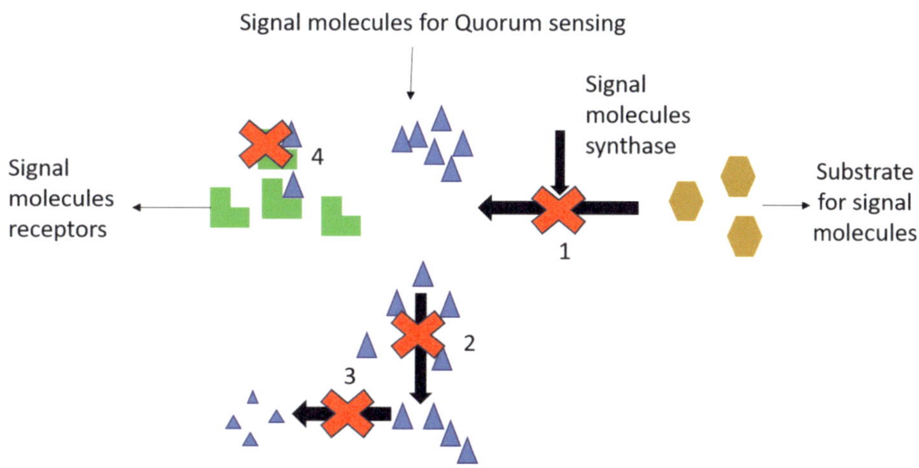

Figure 3.
Quorum sensing inhibition mechanisms includes inhibition of synthesis of signal molecules (1), inhibition of transport of signal molecules (2), their degradation (3), and receptor inhibition of signal molecules through competitive inhibition.

6. Conclusion

As the antibiotic use increased in the world, the issue of their dispersion also increased in the environment resulting in the need for new and broad-spectrum antimicrobial compounds. The world has been trying to overcome the threat since the beginning of the development of resistance in bacteria against powerful antimicrobials, but the issue is not solved fully as it is present in almost all environments, and removal is not possible. The antimicrobial stewardship program has been designed keeping in mind the uncontrolled dispersion of resistance in the environmental compartments. Therapeutic strategies are highly anticipated with modern technology to eradicate multi-drug resistance from the environment and reduce the cost of treatment for the general public. Additionally, a comprehensive understanding of the advantages and disadvantages of the dispersion and enrichment of the ARB and ARGs in the environment and the use of modern and easily accessible technologies could broaden our horizon to get rid of the global threat permanently for all stakeholders.

Author details

Sadia Khan
Department of Environmental Engineering, NED University of Engineering and Technology, Karachi, Pakistan

*Address all correspondence to: sadiakhan@cloud.neduet.edu.pk

References

[1] Harikumar G, Krishanan K. The growing menace of drug resistant pathogens and recent strategies to overcome drug resistance: A review. Journal of King Saud University – Science. 2022;**34**(4):101979. ISSN 1018-3647. DOI: 10.1016/j.jksus.2022.101979

[2] Medina E, Pieper DH. Tackling threats and future problems of multidrug-resistant bacteria. Current Topics in Microbiology and Immunology. 2016;**398**:3-33. DOI: 10.1007/82_2016_492

[3] Wiggins BA, Andrews RW, Conway RA, Corr CL, Dobratz EJ, Dougherty DP, et al. Use of antibiotic resistance analysis to identify nonpoint sources of fecal pollution. Applied and Environmental Microbiology. 1999;**65**(8):3483-3486. DOI: 10.1128/AEM.65.8.3483-3486.1999

[4] Khan S, Beattie TK, Knapp CW. The use of minimum selectable concentrations (MSCs) for determining the selection of antimicrobial resistant bacteria. Ecotoxicology. 2017;**26**(2):283-292. Published online on 7 Feb, 2017. DOI: 10.1007/s10646-017-1762-y

[5] Barancheshme F, Munir M. Strategies to combat antibiotic resistance in the wastewater treatment plants. Frontiers in Microbiology. 2018;**8**:2603. DOI: 10.3389/fmicb.2017.02603

[6] Honda R, Matsuura N, Sorn S, et al. Transition of antimicrobial resistome in wastewater treatment plants: Impact of process configuration, geographical location and season. npj Clean Water. 2023;**6**:46. DOI: 10.1038/s41545-023-00261-x

[7] Yuan X, Cui K, Chen Y, et al. Response of antibiotic and heavy metal resistance genes to the co-occurrence of gadolinium and sulfamethoxazole in activated sludge systems. Frontiers of Environmental Science & Engineering. 2023;**17**:154. DOI: 10.1007/s11783-023-1754-5

[8] Khan S, Beattie TK, Knapp CW. Rapid selection of antimicrobial-resistant bacteria in complex water systems by chlorine and pipe materials. Environmental Chemistry Letters. 2019;**17**:1367-1373. DOI: 10.1007/s10311-019-00867-z

[9] Khan S, Knapp CW, Beattie TK. Antibiotic resistant bacteria found in municipal drinking water. Environmental Processes. 2016;**3**(3):541-552. DOI: 10.1007/s40710-016-0149-z

[10] Hamidin N, Yu QJ, Connell DW. Human health risk assessment of chlorinated disinfection by-products in drinking water using a probabilistic approach. Water Research. 2008;**42**(13):3263-3274. DOI: 10.1016/j.watres.2008.02.029

[11] Richardson SD, Postigo C. Drinking water disinfection by-products. In: Barceló D, editor. Emerging Organic Contaminants and Human Health. The Handbook of Environmental Chemistry. Vol. 20. Berlin, Heidelberg: Springer; 2011. DOI: 10.1007/698_2011_125

[12] Li X-F, Mitch WA. Drinking water disinfection Byproducts (DBPs) and human health effects: Multidisciplinary challenges and opportunities. Environmental Science & Technology. 2018;**52**(4):1681-1689. DOI: 10.1021/acs.est.7b05440

[13] Nguyen C, Elfland C, Edwards M. Impact of advanced water conservation features and new copper pipe

on rapid chloramine decay and microbial regrowth. Water Research. 2012;**46**(3):611-621. DOI: 10.1016/j.watres.2011.11.006

[14] Nguyen CK, Powers KA, Raetz MA, Parks JL, Edwards MA. Rapid free chlorine decay in the presence of Cu(OH)2: Chemistry and practical implications. Water Research. 2011;**45**(16):5302-5312. DOI: 10.1016/j.watres.2011.07.039

[15] Falkinham JO, Pruden A, Edwards M. Opportunistic premise plumbing pathogens: Increasingly important pathogens in drinking water. Pathogens. 2015;**4**(2):373-386. DOI: 10.3390/pathogens4020373

[16] Falkinham JO 3rd. Common features of opportunistic premise plumbing pathogens. International Journal of Environmental Research and Public Health. 2015;**12**(5):4533-4545. DOI: 10.3390/ijerph120504533

[17] Prevots DR, Marras TK. Epidemiology of human pulmonary infection with nontuberculous mycobacteria: A review. Clinics in Chest Medicine. 2015;**36**(1):13-34. DOI: 10.1016/j.ccm.2014.10.002

[18] Lewis AH, Falkinham JO 3rd. Microaerobic growth and anaerobic survival of Mycobacterium avium, mycobacterium intracellulare and mycobacterium scrofulaceum. International Journal of Mycobacteriology. 2015;**4**(1):25-30. DOI: 10.1016/j.ijmyco.2014.11.066

[19] Carlson KM, Boczek LA, Chae S, Ryu H. Legionellosis and recent advances in Technologies for Legionella Control in premise plumbing systems: A review. Water. 2020;**12**(3):676. DOI: 10.3390/w12030676

[20] Leslie E, Hinds J, Hai FI. Causes, factors, and control measures of opportunistic premise plumbing pathogens—A critical review. Applied Sciences. 2021;**11**(10):4474. DOI: 10.3390/app11104474

[21] Fujitani S, Sun HY, Yu VL, Weingarten JA. Pneumonia due to Pseudomonas aeruginosa: part I: epidemiology, clinical diagnosis, and source. Chest. 2011;**139**(4):909-919. DOI: 10.1378/chest.10-0166

[22] Mahapatra A, Padhi N, Mahapatra D, Bhatt M, Sahoo D, Jena S, et al. Study of biofilm in bacteria from water pipelines. Journal of Clinical Diagnostic Research. 2015;**9**(3):DC09-11. DOI: 10.7860/JCDR/2015/12415.5715

[23] Rozej A, Cydzik-Kwiatkowska A, Kowalska B, Kowalski D. Structure and microbial diversity of biofilms on different pipe materials of a model drinking water distribution systems. World Journal of Microbiology and Biotechnology. 2015;**31**(1):37-47. DOI: 10.1007/s11274-014-1761-6

[24] Vital M, Stucki D, Egli T, Hammes F. Evaluating the growth potential of pathogenic bacteria in water. Applied and Environmental Microbiology. 2010;**76**(19):6477-6484. DOI: 10.1128/AEM.00794-10

[25] Hayward C, Ross KE, Brown MH, Bentham R, Whiley H. The presence of opportunistic premise plumbing pathogens in residential buildings: A literature review. Water. 2022;**14**(7):1129. DOI: 10.3390/w14071129

[26] Cycoń M, Mrozik A, Piotrowska-Seget Z. Antibiotics in the soil environment-degradation and their impact on microbial activity and diversity. Frontiers in Microbiology.

2019;**10**:338. DOI: 10.3389/fmicb.2019.00338

[27] Xia Y, Sui Q, Lyu S, Zhao W, Liu J, Cai Z, et al. Municipal solid waste landfills: An underestimated source of pharmaceutical and personal care products in the water environment. Environmental Science & Technology. 2020;**54**(16):9757-9768. DOI: 0.1021/acs.est.0c00565

[28] Bloor MC, Kiryushina A, Kydralieva K, Bondarenko L, Pozdnyakov L, Manucharova N, et al. Divergent effects of antibiotics on plants and microbiota in soils with contrasting humus content. Water, Air, and Soil Pollution. 2021;**232**:518. DOI: 10.1007/s11270-021-05459-8

[29] Kemper N. Veterinary antibiotics in the aquatic and terrestrial environment. Ecological Indicators. 2008;**8**(1):1-13. DOI: 10.1016/j.ecolind.2007.06.002

[30] Anand U, Reddy B, Singh VK, Singh AK, Kesari KK, Tripathi P, et al. Potential environmental and human health risks caused by antibiotic-resistant bacteria (ARB), antibiotic resistance genes (ARGs) and emerging contaminants (ECs) from municipal solid waste (MSW) landfill. Antibiotics (Basel). 2021;**10**(4):374. DOI: 10.3390/antibiotics10040374

[31] Osbiston K, Oxbrough A, Fernández-Martínez LT. Antibiotic resistance levels in soils from urban and rural land uses in Great Britain. Access Microbiology. 2020;**3**(1):acmi000181. DOI: 10.1099/acmi.0.000181

[32] Wang X, Lan B, Fei H, Wang S, Zhu G. Heavy metal could drive co-selection of antibiotic resistance in terrestrial subsurface soils. Journal of Hazardous Materials. 2021;**411**:124848, ISSN 0304-3894. DOI: 10.1016/j.jhazmat.2020.124848

[33] Seiler C, Berendonk TU. Heavy metal driven co-selection of antibiotic resistance in soil and water bodies impacted by agriculture and aquaculture. Frontiers in Microbiology. 2012;**3**:399. DOI: 10.3389/fmicb.2012.00399

[34] Cogliani C, Goossens H, Greko C. Restricting antimicrobial use in food animals: Lessons from Europe. Microbe. 2011;**6**(6):274-279

[35] Kelley P. Antimicrobial stewardship: The role of the clinical microbiology service. Pathology. 2014;**46**(Suppl. 1):S45. ISSN 0031-3025. DOI: 10.1097/01.PAT.0000443495.55603.fb

[36] Gerber JS, Jackson MA, Tamma PD, et al. AAP committee on infectious diseases, pediatric infectious diseases society. Antibiotic Stewardship in Pediatrics. Pediatrics. 2021;**147**(1):e202004029

[37] Huong VTL, Ngan TTD, Thao HP, Quang LM, Hanh TTT, Hien NT, et al. Assessing feasibility of establishing antimicrobial stewardship programmes in two provincial-level hospitals in Vietnam: An implementation research study. BMJ Open. 2021;**11**(10):e053343. DOI: 10.1136/bmjopen-2021-053343

[38] Yang B, Fang D, Lv Q, Wang Z, Liu Y. Targeted therapeutic strategies in the battle against pathogenic bacteria. Frontiers in Pharmacology. 2021;**12**:673239. DOI: 10.3389/fphar.2021.673239

[39] León-Buitimea A, Garza-Cárdenas CR, Garza-Cervantes JA, Lerma-Escalera JA, Morones-Ramírez JR. The demand for new antibiotics: Antimicrobial peptides, nanoparticles, and combinatorial therapies as future

strategies in antibacterial agent design. Frontiers in Microbiology. 2020;**11**:1669. DOI: 10.3389/fmicb.2020.01669

[40] Anyanwu MU, Jaja IF, Nwobi OC. Occurrence and characteristics of Mobile Colistin resistance (mcr) gene-containing isolates from the environment: A review. International Journal of Environmental Research and Public Health. 2020;**17**(3):1028. DOI: 10.3390/ijerph17031028

[41] Wales AD, Davies RH. Co-selection of resistance to antibiotics, biocides and heavy metals, and its relevance to foodborne pathogens. Antibiotics. 2015;**4**:567-604. DOI: 10.3390/antibiotics4040567

[42] Aslam B, Wang W, Arshad MI, Khurshid M, Muzammil S, Rasool MH, et al. Antibiotic resistance: A rundown of a global crisis. Infection and Drug Resistance. 2018;**11**:1645-1658. DOI: 10.2147/IDR.S173867

[43] Principi N, Silvestri E, Esposito S. Advantages and limitations of bacteriophages for the treatment of bacterial infections. Frontiers in Pharmacology. 2019;**10**:513. DOI: 10.3389/fphar.2019.00513

[44] Su Y, Wu D, Xia H, Zhang C, Shi J, Wilkinson KJ, et al. Metallic nanoparticles induced antibiotic resistance genes attenuation of leachate culturable microbiota: The combined roles of growth inhibition, ion dissolution and oxidative stress. Environment International. 2019;**128**:407-416. ISSN 0160-4120. DOI: 10.1016/j.envint.2019.05.007

[45] Khameneh B, Iranshahy M, Soheili V, et al. Review on plant antimicrobials: A mechanistic viewpoint. Antimicrob resist. Infection Control. 2019;**8**:118. DOI: 10.1186/s13756-019-0559-6

[46] Zhao X, Yu Z, Ding T. Quorum-sensing regulation of antimicrobial resistance in bacteria. Microorganisms. 2020;**8**(3):425. DOI: 10.3390/microorganisms8030425

[47] Jingjing Z, Tao F, Jiayi W, et al. The mechanisms and applications of quorum sensing (QS) and quorum quenching (QQ). Journal of Ocean University of China. 2019;**18**(6):1427-1442

[48] Avdic E, Carroll KC. The role of the microbiology laboratory in antimicrobial stewardship programs. Infectious Disease Clinics of North America. 2014;**28**(2):215-235. DOI: 10.1016/j.idc.2014.01.002. Epub 2014 Apr 3

[49] Bouza E, Muñoz P, Burillo A. Role of the clinical microbiology Laboratory in Antimicrobial Stewardship. The Medical Clinics of North America. 2018;**102**(5):883-898. DOI: 10.1016/j.mcna.2018.05.003 Epub 2018 Jul 14

Chapter 10

Antimicrobial Stewardship in the Management of Multidrug-Resistant Gram-Negative Bacteria Infections

Temitope Oyedemi, Tolulope Fadeyi and Kolapo Fasina

Abstract

Antimicrobial resistance constitutes a public health menace globally, affecting treatment outcomes in healthcare settings. This problem is exacerbated in Gram-negative bacteria including *Klebsiella pneumonia*, *Pseudomonas aeruginosa*, *Acinetobacter baumannii*, *E. coli*, *Salmonella* spp., and others belonging to the Enterobacteriaceae family. These organisms have developed resistance mechanisms that render common antibiotics ineffective, making infections caused by these pathogens difficult to treat. Particularly, unregulated antibiotic use, selective pressure, and horizontal gene transfer are some of the contributors to their resistance to the available antibiotics. Effective antimicrobial stewardship plays a crucial role in managing these infections and preventing their further escalation through Antimicrobial Stewardship programs, de-escalation therapy, combination therapy, antibiotics dose optimization, and prophylactic antibiotic are used in those at high risk of infection. Education and training are vital for healthcare providers to enhance their knowledge of antimicrobial stewardship principles and implementation.

Keywords: antimicrobial resistance, Gram-negative bacteria, antibiotics, de-escalation therapy, antimicrobial stewardship

1. Introduction

1.1 Background and significance

Antimicrobial stewardship (AMS) is defined as "an organizational or healthcare-system-wide approach for fostering and monitoring judicious use of antimicrobials to preserve their effectiveness" [1]. Multidrug-resistant Gram-negative bacterial infections have become a major global health concern due to the limited treatment options available and the high mortality rates associated with these infections [2]. Gram-negative bacteria, such as *Escherichia coli*, *Klebsiella pneumoniae*, *Acinetobacter baumannii*, and *Pseudomonas aeruginosa*, have developed resistance mechanisms that render common antibiotics ineffective, making infections caused by these pathogens difficult to treat [3]. Antimicrobial stewardship programs focus on improving the appropriate use of antimicrobial agents, to preserve their effectiveness and minimize the emergence and spread of resistance [4]. In the case of multidrug-resistant Gram-negative bacterial infections, effective antimicrobial stewardship plays a crucial role in managing these infections and preventing their further escalation [5].

The topic of antimicrobial stewardship in the management of multidrug-resistant Gram-negative bacterial infections holds significant importance due to several reasons:

1. *Addressing the global health threat*: Multidrug-resistant Gram-negative bacterial infections pose a significant global health threat, leading to increased morbidity, mortality, and healthcare costs. Antimicrobial stewardship plays a vital role in combating this threat by promoting rational antibiotic use, optimizing treatment regimens, and preventing the further emergence and spread of antimicrobial resistance [6].

2. *Preserving antibiotic effectiveness*: Antimicrobial stewardship programs are crucial for preserving the effectiveness of existing antibiotics. By promoting appropriate antibiotic-prescribing practices, optimizing drug dosage and duration, and reducing unnecessary antibiotic use, antimicrobial stewardship helps slow down the development of resistance and extends the lifespan of available antibiotics [7].

3. *Enhancing patient outcomes*: Multidrug-resistant Gram-negative bacterial infections are associated with poorer patient outcomes, including increased treatment failure rates, prolonged hospital stays, and higher mortality rates. Effective antimicrobial stewardship ensures that patients receive the most appropriate and effective antibiotics, leading to improved clinical outcomes and reduced adverse events [7, 8].

4. *Reducing healthcare costs*: The economic burden associated with multidrug-resistant Gram-negative bacterial infections is significant. Antimicrobial stewardship programs help optimize antibiotic use, prevent unnecessary treatments, and reduce the length of hospital stays, resulting in cost savings for healthcare systems, patients, and society as a whole [9, 10].

5. *Combating antimicrobial resistance*: Antimicrobial stewardship is a crucial component of the global efforts to combat antimicrobial resistance. By implementing evidence-based practices, surveillance of antibiotic use and resistance patterns, and education and training initiatives, antimicrobial stewardship programs contribute to the overall reduction of antimicrobial resistance, protecting the effectiveness of antibiotics for future generations [11, 12].

6. *Improving infection control*: Multidrug-resistant Gram-negative bacteria are often associated with healthcare-associated infections. Antimicrobial stewardship programs emphasize infection prevention and control measures, such as hand hygiene, environmental cleaning, and adherence to best practices, which can help reduce the transmission of these resistant pathogens and prevent outbreaks [13].

7. *Interdisciplinary collaboration*: Antimicrobial stewardship requires collaboration among various healthcare professionals, including physicians, pharmacists, microbiologists, infection control practitioners, and administrators. By fostering

interdisciplinary teamwork and communication, antimicrobial stewardship programs promote a holistic approach to patient care, ensuring that the right antibiotics are prescribed at the right time and in the right dosage [14].

1.2 Multidrug-resistant Gram-negative bacteria (MDR-GNB)

Multidrug-resistant Gram-negative bacteria (MDR-GNB) refer to the ability of these bacteria to exhibit resistance to multiple classes of antibiotics [15, 16]. Gram-negative bacteria are a group of bacteria that have a distinct cell wall structure and include common pathogens such as *Escherichia coli*, *Klebsiella pneumoniae*, *Acinetobacter baumannii*, and *Pseudomonas aeruginosa*. The term "multidrug" signifies resistance to multiple classes of antibiotics, which means that these bacteria are unaffected by several types of antimicrobial agents commonly used in clinical practice. This resistance can include resistance to β-lactams (such as penicillins and cephalosporins), fluoroquinolones, aminoglycosides, carbapenems, and other important antibiotic classes. Multidrug-resistant Gram-negative bacteria are of particular concern because they limit the treatment options available to healthcare professionals, leading to increased reliance on less effective or more toxic antibiotics [17]. MDR-GNB has acquired genetic mechanisms that enable them to evade the effects of different antibiotics, making them difficult to treat. These resistance mechanisms can include the production of enzymes that inactivate antibiotics, changes in the bacterial cell wall structure that prevent drug entry, and efflux pumps that actively remove antibiotics from within the bacterial cell [18, 19].

1.3 Importance of antimicrobial stewardship in managing multidrug-resistant infections

Antimicrobial stewardship plays a critical role in managing multidrug-resistant infections by addressing the challenges posed by antimicrobial resistance. The importance of antimicrobial stewardship in managing multidrug resistance infections can be summarized as follows:

1. *Rational and appropriate antibiotic use*: Antimicrobial stewardship programs promote the rational and appropriate use of antibiotics. Antibiotic use has been associated with the persistence of resistance genes in infants gut [20], thus stressing the importance of regulating antibiotic use. This involves ensuring that antibiotics are prescribed only when necessary, selecting the most effective drug based on local resistance patterns, optimizing dosing regimens, and considering duration of treatment. By using antibiotics judiciously, antimicrobial stewardship helps minimize the selection pressure that drives the development and spread of multidrug-resistant bacteria [11, 21].

2. *Preservation of antibiotic effectiveness*: Multidrug resistance threatens the effectiveness of existing antibiotics. Antimicrobial stewardship aims to preserve the efficacy of available antibiotics by preventing their overuse or misuse [21, 22]. By implementing strategies to reduce unnecessary antibiotic prescription, such as avoiding empirical treatment when it is not warranted and de-escalating therapy based on culture results, antimicrobial stewardship helps maintain the effectiveness of antibiotics for both current and future patients.

3. *Preventing treatment failures and complications*: Multidrug-resistant infections are associated with higher treatment failure rates and increased morbidity and mortality. Antimicrobial stewardship programs ensure that patients receive appropriate and effective antibiotic therapy, tailored to the specific infection and local resistance patterns [7, 23]. This helps minimize treatment failures, reduce the risk of complications, and improve patient outcomes.

4. *Reducing adverse effects and collateral damage*: Certain antibiotics used to treat multidrug-resistant infections can have significant side effects or toxicities. Antimicrobial stewardship helps minimize the exposure of patients to potentially harmful antibiotics by promoting the use of more targeted therapies and avoiding unnecessary or prolonged treatment courses [24]. This reduces the occurrence of adverse effects and associated patient harm.

5. *Controlling the spread of multidrug-resistant bacteria*: Antimicrobial stewardship programs emphasize infection prevention and control measures, such as hand hygiene, isolation precautions, and appropriate environmental cleaning [25]. By effectively implementing these measures, antimicrobial stewardship helps contain the spread of multidrug-resistant bacteria within healthcare settings and reduces the risk of outbreaks.

6. *Optimizing healthcare resource utilization*: Multidrug-resistant infections are often associated with increased healthcare costs, longer hospital stays, and additional resource utilization. Antimicrobial stewardship programs can help optimize resource utilization by reducing the length of hospital stays, minimizing the need for expensive broad-spectrum antibiotics, and preventing complications related to inappropriate antibiotic use [7, 23]. This leads to cost savings for healthcare systems and improved allocation of limited resources.

7. *Combating antimicrobial resistance*: Multidrug resistance is a significant contributor to the global problem of antimicrobial resistance. Antimicrobial stewardship programs play a vital role in addressing this challenge by promoting responsible antibiotic use, reducing the emergence and spread of resistant bacteria, and preserving the effectiveness of antibiotics for future generations [11].

2. Challenges in managing multidrug-resistant Gram-negative bacterial infections

Managing multidrug-resistant Gram-negative (MDR-GNB) bacterial infections poses several challenges, which include:

2.1 Limited treatment options

The challenge of limited treatment options in managing multidrug-resistant Gram-negative bacterial infections refers to the reduced effectiveness of available antibiotics against these infections. MDR-GNB has developed various mechanisms to evade the effects of commonly used antibiotics, making them resistant to multiple classes of drugs. MDR-GNB often exhibit resistance to multiple classes of antibiotics, limiting the available treatment options [26]. This scarcity of effective antibiotics can make it challenging to find appropriate therapies that can effectively target and eliminate the infection. Moreover, MDR-GNB are associated with higher morbidity and mortality rates compared to infections caused by susceptible strains [27]. Treatment failures, prolonged hospital stays, and increased risk of complications contribute to the poorer outcomes observed in patients with these infections.

2.2 Rapid spread of resistance mechanism

Gram-negative bacteria possess complex resistance mechanisms, including the production of enzymes that can degrade antibiotics, altered outer membrane permeability, and efflux pumps that actively remove antibiotics from within the bacterial cell [28, 29]. These complex resistance mechanisms make it difficult to combat multidrug-resistant infections and require the use of multiple strategies to overcome them. Several factors contribute to the complex and rapid spread of resistance mechanisms. These include genetic plasticity, which enables GNB to acquire resistance genes through various mechanisms [30]. These bacteria can acquire resistance genes through horizontal gene transfer, where genetic material is transferred between different bacteria, including those of the same or different species. This transfer can occur through plasmids, integrons, or transposons, facilitating the rapid spread of resistance mechanisms [31]. These elements can move between bacteria, allowing resistance genes to be transferred to new bacteria strains or species. Mobile genetic elements contribute to the horizontal transfer of resistance mechanisms and can rapidly disseminate resistance within a population or across different geographical locations. Likewise, the widespread use and misuse of antibiotics create a selective pressure that favors the survival and proliferation of multidrug-resistant bacteria [32]. Exposure to antibiotics provides a survival advantage to the bacteria carrying resistance mechanisms, allowing them to outcompete susceptible strains. The continuous exposure to suboptimal antibiotic concentrations or incomplete treatment courses can further drive the selection and expansion of resistant strains [33]. Healthcare settings and community transmission also contribute to the spread of MDR-GNB within healthcare facilities, where vulnerable patients with compromised immune systems and invasive medical procedures create opportunities for transmission. The complex network of healthcare-associated infections allows for the rapid dissemination of resistant strains. Additionally, these bacteria can also spread in community settings, where factors such as overcrowding, poor hygiene practices, and close contact contribute to their transmission [34].

2.3 The need for effective infection prevention

Effective infection prevention is crucial in managing multidrug-resistant Gram-negative bacterial infections due to high transmission rates [35]. MDR-GNB has the potential for rapid transmission within healthcare settings and the community [36]. These bacteria can survive on surfaces, medical equipment, and the hands of healthcare workers, facilitating their spread to susceptible individuals [37]. The challenge lies in preventing the transmission of these resistant bacteria to vulnerable patients and limiting the development of outbreaks. MDR-GNB can persist in the environment for extended periods, surviving on surfaces and in water sources. This persistence increases the risk of indirect transmission to patients and poses challenges for effective infection prevention. Rigorous cleaning and disinfection protocols, as well as appropriate water management strategies, are necessary to reduce the environmental reservoir of these resistant bacteria. The ability of MDR-GNB to colonize various sites in the human body, including the skin, gastrointestinal tract, and respiratory tract is another factor [38]. Colonized individuals serve as a reservoir for transmission to others. Addressing the challenge of effective infection prevention requires a comprehensive approach that includes strict adherence to infection control measures, robust environmental cleaning and disinfection practices, surveillance of multidrug-resistant organisms, and appropriate use of antimicrobials. It also involves active collaboration between healthcare facilities, public health agencies, and the communities to raise awareness, implement preventive measures, and reduce the burden of multidrug-resistant Gam-negative bacterial infections.

3. Antimicrobial stewardship strategies for managing multidrug resistance

3.1 Antimicrobial stewardship programs (ASPs)

Antimicrobial Stewardship Programs (ASPs) are comprehensive strategies designed to optimize the use of antimicrobial agents in healthcare settings. They are crucial in managing multidrug-resistant infections. They help to preserve the effectiveness of existing antibiotics, reduce treatment failures, prevent the emergence of resistance, improve patient outcomes, and optimize healthcare resource utilization. Additionally, ASPs contribute to the global effort of combating antimicrobial resistance by promoting responsible antibiotic use and reducing the spread of multidrug-resistant bacteria [7, 9, 11].

3.2 De-escalation therapy

De-escalation therapy is a strategy employed in the management of multidrug resistance that involves starting empiric broad-spectrum antibiotic treatment and subsequently narrowing down the antibiotic regimen based on clinical response and susceptibility testing. The goal of de-escalation is to optimize antimicrobial therapy by reducing the use of broad-spectrum antibiotics and minimizing the selection pressure for antibiotic resistance [39, 40]. De-escalation therapy begins with the administration of empiric broad-spectrum antibiotics that cover a wide range of potential pathogens, including multidrug-resistant bacteria. This initial approach ensures that appropriate antimicrobial therapy is promptly initiated, particularly in cases of severe infections where the causative organism may not be immediately identifiable. During de-escalation therapy, close monitoring of the patient's clinical response is essential. If the patient shows signs of improvement, such as decreased fever, resolution of symptoms, and improvement in laboratory markers, it indicates that the initial empiric therapy is effective [41]. Additionally, microbiological assessment, including culture and susceptibility testing, is crucial in identifying the causative pathogen and determining its susceptibility to antibiotics. Once the microbiological results become available, the antibiotic regimen is tailored to the specific pathogen and its susceptibility profile. De-escalation involves narrowing down the antibiotic coverage by selecting a more targeted and narrow-spectrum antibiotic or combination of antibiotics [42]. This approach helps minimize the unnecessary use of broad-spectrum antibiotics, which are associated with a higher risk of promoting antimicrobial resistance. De-escalation therapy takes into account the local resistance patterns and susceptibility data of pathogens commonly encountered in the healthcare facility or community. By considering local epidemiology, clinicians can choose antibiotics that are effective against the prevalent multidrug-resistant bacteria while avoiding unnecessary use of broad-spectrum agents [43].

3.3 Combination therapy

Combination therapy, as a strategy for managing multidrug resistance, involves the simultaneous use of two or more antibiotics with different mechanisms of action to enhance treatment efficacy and overcome resistance mechanisms. Combination therapy aims to achieve synergistic effects, where the combined action of multiple antibiotics is greater than the sum of their individual effects [44]. By targeting different bacterial pathways or mechanisms, combination therapy can enhance bacterial killing, inhibit resistance mechanisms, and improve treatment outcomes. Combination therapy allows for broader coverage of potential pathogens, including those that may exhibit resistance to individual antibiotics. By combining antibiotics with different spectra of activity, the regimen can effectively target a wider range of bacteria, including multidrug-resistant strains. This broader coverage increases the likelihood of successful treatment, especially in infections where the causative organisms are unknown or resistant pathogens are prevalent. The use of combination therapy can help prevent the emergence of resistance [45]. Resistance mechanisms can be complex and multifaceted, requiring the coordinated action of multiple genetic changes in bacteria. By utilizing multiple antibiotics that target different pathways, the likelihood of bacteria developing simultaneous resistance to all components of the combination is reduced. This reduces the risk of treatment failure due to the emergence of resistance during therapy. Combination therapy can overcome existing resistance in multidrug-resistant bacteria. By utilizing antibiotics with different mechanisms of action, combination therapy can circumvent specific resistance mechanisms employed by bacteria. For example, if a bacterium possesses efflux pumps that can actively pump out a particular antibiotic, combining it with another antibiotic that is not a substrate for the efflux pump can help overcome this resistance mechanism and improve treatment efficacy [46].

3.4 Dose optimization

Dose optimization is a strategy employed in the management of multidrug resistance that focuses on achieving optimal drug concentrations at the site of infection to maximize treatment efficacy against resistant pathogens [47]. By adjusting the dosage of antibiotics, dose optimization aims to overcome resistance mechanisms, improve bacterial killing, and enhance treatment outcomes. Dose optimization takes into account the pharmacokinetics (PK) and pharmacodynamics (PD) of antibiotics [48]. PK refers to how the body processes a drug, including absorption, distribution, metabolism, and elimination. PD refers to the relationship between drug exposure and its effect on bacteria. Understanding the PK/PD properties of antibiotics is crucial in determining the optimal dosage regimen to achieve effective drug concentrations at the site of infection. The effectiveness of antibiotics against resistant pathogens is often related to achieving adequate drug exposure. By optimizing the dosage, the goal is to ensure that the antibiotic concentration at the site of infection exceeds the minimum inhibitory concentration (MIC) for the target pathogen. This exposure-response relationship is critical in achieving optimal bacterial killing and minimizing the risk of treatment failure [49]. Dose optimization should be employed alongside other comprehensive strategies and guided by PK/PD principles and individual patient factors to ensure the effective management of multidrug resistance.

3.5 Prophylactic antibiotic use

Prophylactic antibiotic use, as a strategy for managing multidrug resistance, involves the administration of antibiotics to prevent infections in individuals at high risk of developing infections [50]. While the primary aim of prophylaxis is to prevent infections, it can also play a role in managing multidrug resistance. Prophylactic antibiotics are typically administered to individuals who are at increased risk of developing infections, such as surgical patients, immunocompromised individuals, or patients with specific medical conditions [51]. By targeting those at high risk, prophylaxis aims to prevent infections that could potentially lead to multidrug-resistant bacteria colonization or subsequent spread. The choice of antibiotics for prophylaxis is important to consider. Narrow-spectrum antibiotics with activity against the expected pathogens in the specific procedure or patient population are preferred. This approach helps minimize the disruption of the normal flora and reduces the selective pressure for multidrug resistance. Additionally, the selection of antibiotics should consider local resistance patterns and susceptibility data to ensure their effectiveness. The duration of prophylactic antibiotic use should be limited to the perioperative or high-risk period to avoid unnecessary prolonged exposure. Prolonged prophylactic use can lead to increased antibiotic consumption, potential adverse effects, and the development of resistance [52]. It is essential to adhere to recommended guidelines regarding the appropriate duration of prophylaxis for each indication. Prophylactic antibiotic use should be part of a comprehensive approach to infection prevention that includes other measures such as strict aseptic techniques, proper hand hygiene, adherence to infection control protocols, and vaccination. Combining prophylactic antibiotics with these multimodal strategies can further reduce the risk of infections and multidrug resistance. However, the risk of promoting antibiotic resistance should always be considered, and prophylactic antibiotic use should be tailored to individual patient needs while emphasizing the importance of antimicrobial stewardship.

3.6 Guidelines on antibiotic prescription

Guidelines on antibiotic prescription are an essential strategy for managing multidrug-resistant bacteria by promoting appropriate and judicious use of antibiotics. These guidelines provide evidence-based recommendations to healthcare providers regarding the selection, dosage, and duration of antibiotic therapy for specific infections [53]. Guidelines on antibiotic prescribing promote rational antibiotic use by guiding healthcare providers in choosing the most effective antibiotic therapy while minimizing the risk of resistance [11, 54]. They provide recommendations based on the local epidemiology of bacterial pathogens and their susceptibility patterns, considering factors such as resistance rates, healthcare-associated infections, and community-acquired infections. Guidelines take into account the specific characteristics of different infections and provide recommendations for tailored treatment [55]. By providing guidance on appropriate antibiotic selection, dosage, and duration, guidelines help ensure that patients receive optimal therapy, increasing the likelihood of successful treatment outcomes and minimizing the development of multidrug resistance. It also empowers healthcare providers with up-to-date information on the latest advancements in antimicrobial therapy and resistance patterns [56, 57].

4. The roles of healthcare providers in antimicrobial stewardship

4.1 Multidisciplinary team approach

The multidisciplinary team approach is a crucial aspect of antimicrobial stewardship, involving collaboration among various healthcare providers with different areas of expertise [58]. This team typically includes physicians, pharmacists, microbiologists, infection control practitioners, and other relevant healthcare professionals. The multidisciplinary team brings together professionals from different backgrounds, each with their unique knowledge and perspectives on antimicrobial use. Physicians provide clinical expertise; pharmacists contribute their knowledge of drug interactions and dosing; microbiologists offer insights into microbial identification and susceptibility, and infection control practitioners focus on preventing healthcare-associated infections. By working collaboratively, the team can develop comprehensive strategies to optimize antimicrobial use. Multidisciplinary teams play a crucial role in developing and implementing evidence-based guidelines and protocols for antimicrobial use. These guidelines provide standardized approaches to the diagnosis, treatment, and monitoring of infections, considering local epidemiology and antimicrobial resistance patterns. The team members contribute their expertise to ensure that guidelines are practical, feasible, and align with the best available evidence. The multidisciplinary team approach allows for continuous quality improvement in antimicrobial stewardship [59].

4.2 Education and training

Education and training are essential roles of healthcare providers in antimicrobial stewardship. Education plays a vital role in raising awareness about antimicrobial resistance and the importance of antimicrobial stewardship [60]. Healthcare providers need to understand the consequences of inappropriate antimicrobial use, including the development of resistance, increased healthcare costs, and adverse patient outcomes. Training programs provide healthcare professionals with updated information on antimicrobial resistance patterns, new treatment guidelines, and emerging resistance mechanisms. By staying informed, healthcare providers can make informed decisions and contribute to effective stewardship practices. Education and training initiatives emphasize the importance of patient education regarding antimicrobial use. Healthcare providers learn how to effectively communicate with patients, explaining when antibiotics are necessary and educating them about the risks of inappropriate antibiotic use. Patients are educated about the importance of completing the full course of antibiotics as prescribed, avoiding the sharing or saving of antibiotics, and understanding that antibiotics are not effective against viral infections. By engaging patients in these discussions, healthcare providers help reinforce responsible antibiotic use and promote patient adherence [61].

4.3 Prescription and dispensing practices

Prescription and dispensing practices are vital roles of healthcare providers in antimicrobial stewardship. By following appropriate prescribing and dispensing practices, healthcare providers contribute to the responsible use of antimicrobial agents [62]. Healthcare providers have a responsibility to prescribe antimicrobial agents judiciously. By following evidence-based guidelines and considering factors such as the type and severity of the infection, local resistance patterns, and patient-specific factors (e.g., allergies and renal function), healthcare providers can select the most appropriate antibiotic therapy. This ensures that antibiotics are used only when necessary and minimizes the risk of antibiotic resistance. When prescribing antibiotics, healthcare providers should consider using narrow-spectrum antibiotics whenever possible. Narrow-spectrum antibiotics target specific bacteria, reducing the disruption of the body's normal flora and minimizing the selection pressure for resistance. Targeted therapy, based on culture and susceptibility results, helps ensure that the chosen antibiotic is effective against the identified pathogen [63]. This approach minimizes the unnecessary use of broad-spectrum antibiotics, which can contribute to the emergence of multidrug-resistant organisms.

4.4 Use of clinical decision support systems

The use of Clinical Decision Support Systems (CDSS) is an important role of healthcare providers in antimicrobial stewardship. CDSS refers to computer-based tools that provide healthcare professionals with clinical knowledge and patient-specific information to assist in decision-making [64]. CDSS can integrate evidence-based guidelines for antimicrobial use directly into the clinical workflow. This allows healthcare providers to access up-to-date recommendations at the point of care. The CDSS can provide guidance on appropriate antibiotic selection, dosing, and duration of therapy based on the specific infection and patient characteristics. By incorporating guidelines, CDSS helps ensure that healthcare providers have easy access to the most current and relevant information to support their prescribing decisions. CDSS can provide real-time alerts and reminders to healthcare providers when prescribing antimicrobials. These alerts can notify providers about potential drug interactions, allergies, duplicate therapies, or inappropriate antibiotic choices. By receiving these alerts, healthcare providers can review and reconsider their prescribing decisions, potentially leading to more appropriate antibiotic selection and improved patient outcomes [65]. CDSS can integrate with microbiology laboratory systems to provide real-time access to microbial identification and susceptibility results. This allows healthcare providers to make informed decisions regarding appropriate antibiotic therapy. CDSS can also incorporate local antimicrobial resistance data, providing information on prevalent resistance patterns and suggesting alternative antibiotics when necessary. By utilizing these integrated data sources, healthcare providers can make more targeted antibiotic choices and avoid unnecessary broad-spectrum use.

5. Best practices in antimicrobial stewardship in multidrug-resistant Gram-negative bacterial infections

5.1 Surveillance and monitoring

Surveillance and monitoring are crucial best practices in antimicrobial stewardship, particularly when addressing multidrug-resistant Gram-negative bacterial infections [66]. Surveillance and monitoring systems allow healthcare facilities to identify and track the emergence and spread of MDR-GNB. By routinely collecting and analyzing data on bacterial isolates and their antibiotic susceptibility patterns, healthcare providers can detect changes in resistance profiles and identify emerging resistance mechanisms [67]. This early detection helps guide appropriate empiric antibiotic therapy and infection control measures. Surveillance data on MDR-GNB infections enable healthcare providers to make informed decisions regarding empiric antibiotic therapy. Knowing the local resistance patterns helps in selecting antibiotics that are likely to be effective against the prevalent pathogens. This targeted approach improves patient outcomes by reducing the risk of inadequate initial therapy and minimizing the unnecessary use of broad-spectrum antibiotics. Surveillance systems enable the monitoring of trends and changes in antimicrobial resistance patterns over time [68]. By tracking resistance data, healthcare providers can identify shifts in susceptibility profiles, new mechanisms of resistance, or the emergence of novel resistant strains. This information is crucial for guiding treatment decisions, developing local antimicrobial guidelines, and informing public health efforts to address antimicrobial resistance at a regional or national level [69].

Surveillance and monitoring systems enhance the ability to anticipate and respond to new threats posed by MDR-GNB by continuously monitoring resistance patterns, healthcare providers can detect the emergence of resistance to specific antibiotics or classes of antibiotics [70]. This information helps guide the development of alternative treatment strategies, such as novel antimicrobial agents or combination therapies, before resistance becomes widespread. Surveillance and monitoring efforts promote collaboration and data sharing among healthcare facilities, public health agencies, and research institutions. By sharing data on resistance patterns, outbreaks, and best practices, healthcare providers can learn from each other's experiences and adopt successful strategies in their own settings. Collaboration and data sharing also facilitate regional or national surveillance networks, enabling a broader understanding of the epidemiology of multidrug-resistant Gram-negative bacterial infections and supporting coordinated efforts to combat antimicrobial resistance [71].

5.2 Guidelines for antibiotic use

Guidelines for antibiotic use are essential best practices in antimicrobial stewardship, particularly in addressing multidrug-resistant Gram-negative bacterial infections. These guidelines provide evidence-based recommendations for healthcare providers to optimize antimicrobial therapy and combat the challenges posed by resistant bacteria [72]. Guidelines for antibiotic use provide evidence-based recommendations for the selection, dosing, and duration of antimicrobial therapy for multidrug-resistant Gram-negative bacterial infections. These guidelines are developed based on extensive research, clinical trials, and expert consensus. By following these recommendations, healthcare providers can make informed decisions about appropriate antibiotic therapy, considering factors such as the type of infection, local resistance patterns, and individual patient characteristics. Guidelines for antibiotic use emphasize the importance of targeted therapy for multidrug-resistant Gram-negative bacterial infections. Targeted therapy involves selecting antibiotics that are specifically effective against the identified pathogens, taking into account their susceptibility patterns. This approach helps minimize the unnecessary use of broad-spectrum antibiotics, reducing the selective pressure for resistance and preserving the effectiveness of antibiotics. Guidelines may address the use of combination therapy for multidrug-resistant Gram-negative bacterial infections and provide recommendations on when combination therapy may be warranted and which specific combinations are supported by evidence. Guidelines for antibiotic use often include recommendations for implementing infection control measures to prevent the spread of MDR-GNB [66]. These measures may include strategies such as contact precautions, hand hygiene, environmental cleaning, and patient cohort. By incorporating infection control guidelines into antibiotic use recommendations, healthcare providers can contribute to reducing the transmission of resistant bacteria and protecting vulnerable patients.

5.3 Close collaboration between clinicians and microbiologists

Close collaboration between clinicians and microbiologists is considered a best practice in antimicrobial stewardship, particularly in the management of multidrug-resistant Gram-negative bacterial infections [73]. This collaboration facilitates effective communication, decision-making, and patient care. Microbiologists play a critical role in identifying the causative pathogens and their antimicrobial susceptibility patterns. Close collaboration between clinicians and microbiologists ensures that accurate and timely microbiological results are provided to guide appropriate antibiotic therapy. Rapid identification of MDR-GNB allows clinicians to initiate targeted therapy promptly, leading to better patient outcomes and reduced spread of resistant strains. In cases where microbiological results are not immediately available, clinicians rely on empirical antibiotic therapy. Collaboration between clinicians and microbiologists helps in selecting appropriate empiric therapy based on local resistance patterns and the knowledge of circulating MDR-GNB [65]. Microbiologists can provide valuable input regarding the likelihood of resistance to specific antibiotics, helping clinicians make informed decisions and avoid the unnecessary use of broad-spectrum agents. Close collaboration allows for effective feedback and quality improvement initiatives. Microbiologists can provide feedback to clinicians regarding the appropriateness of antibiotic prescribing based on microbiological data. This feedback helps clinicians understand the impact of their prescribing decisions on resistance patterns and guides them in optimizing antibiotic use. Regular meetings and discussions between clinicians and microbiologists enable the exchange of information and the identification of areas for improvement in antimicrobial stewardship practices.

5.4 Benchmarking and quality improvement

Benchmarking and quality improvement are crucial best practices in antimicrobial stewardship, particularly in the management of multidrug-resistant Gram-negative bacterial infections [74]. These practices involve measuring performance, comparing it against established benchmarks or standards, and implementing strategies to improve patient care and outcomes. Benchmarking in antimicrobial stewardship involves defining performance metrics to assess the appropriateness of antibiotic-prescribing practices. These metrics may include indicators such as antibiotic utilization rates, adherence to guidelines, rates of inappropriate prescribing, and outcomes of antimicrobial therapy. By establishing these metrics, healthcare facilities can measure their performance and identify areas that require improvement. Benchmarking allows healthcare facilities to compare their performance against established benchmarks or standards. These benchmarks may be developed based on national guidelines, expert consensus, or data from peer institutions. By comparing their performance, healthcare facilities can identify variations, gaps, or opportunities for improvement in antimicrobial-prescribing practices. Benchmarking facilitates the identification of variations in antimicrobial-prescribing practices among different healthcare providers, departments, or units within a facility [75]. By analyzing these variations, healthcare facilities can identify the best practices and areas where improvement is needed. Understanding the factors contributing to variations in prescribing practices helps develop targeted interventions and share successful strategies across the organization. Benchmarking results guide the implementation of targeted interventions to improve antimicrobial prescribing practices. These interventions may include educational initiatives, clinical decision support systems, antibiotic stewardship rounds, audit and feedback, and guideline development or revision. By tailoring interventions to address specific areas of improvement identified through benchmarking, healthcare facilities can optimize antibiotic use, reduce inappropriate prescribing, and mitigate the development of resistance.

6. Implications for practice

Antimicrobial stewardship has significant implications for practice across healthcare settings. Implementing effective antimicrobial stewardship practices can improve patient outcomes, reduce healthcare costs, and contribute to the global efforts to combat antimicrobial resistance. Antimicrobial stewardship requires strong leadership and support from healthcare administrators and management. Establishing dedicated antimicrobial stewardship teams, providing necessary resources, and fostering a culture that prioritizes responsible antibiotic use are critical for successful implementation and sustainability of antimicrobial stewardship practices.

7. Future directions for research

The field of antimicrobial stewardship continues to evolve, and there are several directions for future research that can further enhance our understanding and implementation of effective stewardship practices. Here are some potential areas for future research:

7.1 Outcome measures and impact assessment

There is a need for standardized and validated outcome measures to assess the impact of antimicrobial stewardship interventions. Future research should focus on developing robust methods to evaluate the clinical, economic, and patient-centered outcomes associated with various stewardship strategies. This will help determine the most effective interventions and their impact on patient outcomes and antimicrobial resistance rates.

7.2 Novel approaches and technologies

Research can explore the development and implementation of innovative approaches and technologies in antimicrobial stewardship. This may include the use of artificial intelligence and machine learning algorithms to guide antibiotic prescription, the application of rapid diagnostic tests to facilitate targeted therapy, or the integration of electronic health records and clinical decision support systems to improve decision-making and optimize antimicrobial use.

7.3 Antimicrobial stewardship in special populations

Future research should focus on understanding the unique challenges and considerations for antimicrobial stewardship in special populations such as pediatrics, elderly patients, immunocompromised individuals, and those with specific comorbidities. Tailored stewardship strategies that address the specific needs and risks in these populations can optimize antibiotic use and improve patient outcomes.

7.4 Antimicrobial stewardship in community settings

While much of the existing research has focused on antimicrobial stewardship in hospitals and healthcare facilities, there is a need for more research on stewardship in community settings. Investigating the effectiveness of interventions, such as education campaigns, public awareness programs, and implementation of guidelines in primary care and outpatient settings, can help promote responsible antibiotic use and reduce the development of resistance.

7.5 One Health approach

Antimicrobial resistance is a complex issue that requires a One Health approach, considering the interplay between human, animal, and environmental factors. Future research can explore the impact of antimicrobial stewardship interventions in veterinary medicine, agricultural practices, and environmental reservoirs of resistance. Understanding the interconnected nature of antimicrobial resistance across different sectors can guide comprehensive and collaborative approaches to stewardship.

7.6 Implementation science and strategies

Research should focus on implementation science and strategies to enhance the adoption and sustainability of antimicrobial stewardship programs. Identifying barriers and facilitators to implementation, understanding the organizational and cultural factors that influence uptake, and developing effective implementation strategies tailored to different healthcare settings can optimize the success and impact of stewardship interventions.

8. Conclusion

Antimicrobial stewardship plays a crucial role in managing multidrug-resistant bacterial infections. De-escalation therapy, combination therapy, dose optimization, and prophylactic antibiotic use are key strategies. Guidelines on antibiotic prescribing, a multidisciplinary team approach, education and training, surveillance and monitoring, close collaboration between clinicians and microbiologists, and benchmarking and quality improvement all contribute to the uptake and performance evaluation of antimicrobial stewardship.

Conflict of interest

The authors declare that there is no conflict of interest.

Author details

Temitope Oyedemi[1*], Tolulope Fadeyi[2] and Kolapo Fasina[3]

1 Department of Microbiology, Adeleke University, Ede, Osun State, Nigeria

2 Department of Pharmaceutical Microbiology, University of Ibadan, Ibadan, Oyo State, Nigeria

3 Biotechnology Division, Rubber Research Institute of Nigeria, Benin, Edo State, Nigeria

*Address all correspondence to: topeoyedemi07@gmail.com

References

[1] Perez KK, Olsen RJ, Musick WL, Cernoch PL, Davis JR, Land GA, et al. Integrating rapid pathogen identification and antimicrobial stewardship significantly decreases hospital costs. Archives of Pathology and Laboratory Medicine. 2013;**137**(9):1247-1254

[2] Cerceo E, Deitelzweig SB, Sherman BM, Amin AN. Multidrug-resistant Gram-negative bacterial infections in the hospital setting: Overview, implications for clinical practice, and emerging treatment options. Microbial Drug Resistance. 2016;**22**(5):412-431

[3] Zagaliotis P, Michalik-Provasek J, Gill JJ, Walsh TJ. Therapeutic bacteriophages for Gram-negative bacterial infections in animals and humans. Pathogens and Immunity. 2022;7(2):1

[4] Razzaque MS. Implementation of antimicrobial stewardship to reduce antimicrobial drug resistance. Expert Review of Anti-infective Therapy. 2021;**19**(5):559-562

[5] Ruiz-Ramos J, Ramírez P. Antimicrobial stewardship programs in the Intensive Care Unit in patients with infections caused by multidrug-resistant Gram-negative bacilli. Medicina Intensiva (English Edition). 2022;**47**(2):99-107

[6] World Health Organization. Antimicrobial stewardship programmes in health-care facilities in low-and middle-income countries. A Practical Toolkit. 2019;**1**:1-71

[7] Al-Omari A, Al Mutair A, Alhumaid S, Salih S, Alanazi A, Albarsan H, et al. The impact of antimicrobial stewardship program implementation at four tertiary private hospitals: Results of a five-years pre-post analysis. Antimicrobial Resistance & Infection Control. 2020;**9**(1):1-9

[8] File TM Jr, Srinivasan A, Bartlett JG. Antimicrobial stewardship: Importance for patient and public health. Clinical Infectious Diseases. 2014;**59** (Suppl. 3):S93-S96

[9] Nathwani D, Varghese D, Stephens J, Ansari W, Martin S, Charbonneau C. Value of hospital antimicrobial stewardship programs [ASPs]: A systematic review. Antimicrobial Resistance & Infection Control. 2019;**8**:1-3

[10] Aiesh BM, Nazzal MA, Abdelhaq AI, Abutaha SA, Zyoud SE, Sabateen A. Impact of an antibiotic stewardship program on antibiotic utilization, bacterial susceptibilities, and cost of antibiotics. Scientific Reports. 2023;**13**(1):5040

[11] Majumder MA, Rahman S, Cohall D, Bharatha A, Singh K, Haque M, et al. Antimicrobial stewardship: Fighting antimicrobial resistance and protecting global public health. Infection and Drug Resistance. 2020;**13**:4713-4738

[12] Hayes JF. Fighting back against antimicrobial resistance with comprehensive policy and education: A narrative review. Antibiotics. 2022;**11**(5):644

[13] Knobloch MJ, McKinley L, Keating J, Safdar N. Integrating antibiotic stewardship and infection prevention and control programs using a team science approach. American Journal of Infection Control. 2021;**49**(8):1072-1074

[14] Sheikh F, Zarowitz BJ, Sheikh F, Naqvi FA, Heil E, Brandt N.

Interprofessional collaboration improves antibiotic stewardship. Journal of the American Medical Directors Association. 2019;**20**(3):B22

[15] Exner M, Bhattacharya S, Christiansen B, Gebel J, Goroncy-Bermes P, Hartemann P, et al. Antibiotic resistance: What is so special about multidrug-resistant Gram-negative bacteria? GMS Hygiene and Infection Control. 2017;**12**:1-24

[16] Bassetti M, Kanj SS, Kiratisin P, Rodrigues C, Van Duin D, Villegas MV, et al. Early appropriate diagnostics and treatment of MDR Gram-negative infections. JAC-Antimicrobial Resistance. 2022;**4**(5):dlac089

[17] Breijyeh Z, Jubeh B, Karaman R. Resistance of Gram-negative bacteria to current antibacterial agents and approaches to resolve it. Molecules. 2020;**25**(6):1340

[18] Reygaert WC. An overview of the antimicrobial resistance mechanisms of bacteria. AIMS Microbiology. 2018;**4**(3):482

[19] Uddin TM, Chakraborty AJ, Khusro A, Zidan BR, Mitra S, Emran TB, et al. Antibiotic resistance in microbes: History, mechanisms, therapeutic strategies and future prospects. Journal of Infection and Public Health. 2021;**14**(12):1750-1766

[20] Fadeyi TE, Oyedemi OT, Awe OO, Ayeni F. Antibiotic use in infants within the first year of life is associated with the appearance of antibiotic-resistant genes in their feces. PeerJ. 2023;**11**:1-16

[21] Wenzler E, Maximos M, Asempa TE, Biehle L, Schuetz AN, Hirsch EB. Antimicrobial susceptibility testing: An updated primer for clinicians in the era of antimicrobial resistance: Insights from the Society of Infectious Diseases Pharmacists. Pharmacotherapy: The Journal of Human Pharmacology and Drug Therapy. 2023;**43**(4):264-278

[22] Ohl CA, Luther VP. Antimicrobial stewardship for inpatient facilities. Journal of Hospital Medicine. 2011;**6**(S1):S4-S15

[23] Barlam TF, Cosgrove SE, Abbo LM, MacDougall C, Schuetz AN, Septimus EJ, et al. Implementing an antibiotic stewardship program: Guidelines by the Infectious Diseases Society of America and the Society for Healthcare Epidemiology of America. Clinical Infectious Diseases. 2016;**62**(10):e51-e77

[24] Lipsky BA, Dryden M, Gottrup F, Nathwani D, Seaton RA, Stryja J. Antimicrobial stewardship in wound care: A position paper from the British Society for Antimicrobial Chemotherapy and European Wound Management Association. Journal of Antimicrobial Chemotherapy. 2016;**71**(11):3026-3035

[25] Taplitz RA, Ritter ML, Torriani FJ. Infection prevention and control, and antimicrobial stewardship. Infectious Diseases. 2017:54

[26] Bassetti M, Peghin M, Vena A, Giacobbe DR. Treatment of infections due to MDR Gram-negative bacteria. Frontiers in Medicine. 2019;**6**:74

[27] Lin TL, Chang PH, Chen IL, Lai WH, Chen YJ, Li WF, et al. Risk factors and mortality associated with multi-drug-resistant Gram-negative bacterial infection in adult patients following abdominal surgery. Journal of Hospital Infection. 2022;**119**:22-32

[28] Li XZ, Plésiat P, Nikaido H. The challenge of efflux-mediated antibiotic

resistance in Gram-negative bacteria. Clinical Microbiology Reviews. 2015;**28**(2):337-418

[29] Darby EM, Trampari E, Siasat P, Gaya MS, Alav I, Webber MA, et al. Molecular mechanisms of antibiotic resistance revisited. Nature Reviews Microbiology. 2023;**21**(5):280-295

[30] Munita JM, Arias CA. Mechanisms of antibiotic resistance. Virulence Mechanisms of Bacterial Pathogens. 2016;**4**(2):481-511

[31] Jian Z, Zeng L, Xu T, Sun S, Yan S, Yang L, et al. Antibiotic resistance genes in bacteria: Occurrence, spread, and control. Journal of Basic Microbiology. 2021;**61**(12):1049-1070

[32] Serwecińska L. Antimicrobials and antibiotic-resistant bacteria: A risk to the environment and to public health. Water. 2020;**12**(12):3313

[33] Ventola CL. The antibiotic resistance crisis: Part 1: Causes and threats. Pharmacy and Therapeutics. 2015;**40**(4):277

[34] Chia PY, Sengupta S, Kukreja A, Ponnampalavanar SSL, Ng OT, Marimuthu K. The role of hospital environment in transmissions of multidrug-resistant Gram-negative organisms. Antimicrobial Resistance & Infection Control. 2020;**9**(29):1-11

[35] Tacconelli EC, Cataldo MA, Dancer SJ, De Angelis G, Falcone M, Frank U, et al. ESCMID guidelines for the management of the infection control measures to reduce transmission of multidrug-resistant Gram-negative bacteria in hospitalized patients. Clinical Microbiology and Infection. 2014;**20**:1-55

[36] Morris S, Cerceo E. Trends, epidemiology, and management of multi-drug resistant Gram-negative bacterial infections in the hospitalized setting. Antibiotics. 2020;**9**(4):196

[37] Suleyman G, Alangaden G, Bardossy AC. The role of environmental contamination in the transmission of nosocomial pathogens and healthcare-associated infections. Current Infectious Disease Reports. 2018;**20**(12):1-11

[38] Moolchandani K, Sastry AS, Deepashree R, Sistla S, Harish BN, Mandal J. Antimicrobial resistance surveillance among intensive care units of a tertiary care hospital in southern India. Journal of Clinical and Diagnostic Research: JCDR. 2017;**11**(2):DC01

[39] Shime N, Kosaka T, Fujita N. De-escalation of antimicrobial therapy for bacteraemia due to difficult-to-treat Gram-negative bacilli. Infection. 2013;**41**:203-210

[40] Lanckohr C, Bracht H. Antimicrobial stewardship. Current Opinion in Critical Care. 2022;**28**(5):551-556

[41] Bassetti S, Tschudin-Sutter S, Egli A, Osthoff M. Optimizing antibiotic therapies to reduce the risk of bacterial resistance. European Journal of Internal Medicine. 2022;**99**:7-12

[42] Mathieu C, Pastene B, Cassir N, Martin-Loeches I, Leone M. Efficacy and safety of antimicrobial de-escalation as a clinical strategy. Expert Review of Anti-Infective Therapy. 2019;**17**(2):79-88

[43] Al Salman J, Al Dabal L, Bassetti M, Alfouzan WA, Al Maslamani M, Alraddadi B, et al. Management of infections caused by WHO critical priority Gram-negative pathogens in Arab countries of the middle east: A consensus paper.

International Journal of Antimicrobial Agents. 2020;**56**(4):106104

[44] Ahmed A, Azim A, Gurjar M, Baronia AK. Current concepts in combination antibiotic therapy for critically ill patients. Indian Journal of Critical Care Medicine: Peer-Reviewed, Official Publication of Indian Society of Critical Care Medicine. 2014;**18**(5):310

[45] Fischbach MA. Combination therapies for combating antimicrobial resistance. Current Opinion in Microbiology. 2011;**14**(5):519-523

[46] Tamma PD, Cosgrove SE, Maragakis LL. Combination therapy for treatment of infections with Gram-negative bacteria. Clinical Microbiology Reviews. 2012;**25**(3):450-470

[47] Póvoa P, Moniz P, Pereira JG, Coelho L. Optimizing antimicrobial drug dosing in critically ill patients. Microorganisms. 2021;**9**(7):1401

[48] Phe K, Heil EL, Tam VH. Optimizing pharmacokinetics-pharmacodynamics of antimicrobial management in patients with sepsis: A review. The Journal of Infectious Diseases. 2020;**222**(Suppl. 2):S132-S141

[49] Drusano GL. Antimicrobial pharmacodynamics: Critical interactions of 'bug and drug'. Nature Reviews Microbiology. 2004;**2**(4):289-300

[50] Aguado JM, Silva JT, Fernández-Ruiz M, Cordero E, Fortún J, Gudiol C, et al. Management of multidrug resistant Gram-negative bacilli infections in solid organ transplant recipients: SET/GESITRA-SEIMC/REIPI recommendations. Transplantation Reviews. 2018;**32**(1):36-57

[51] Buchan CA, Kotton CN, AST Infectious Diseases Community of Practice. Travel medicine, transplant tourism, and the solid organ transplant recipient—guidelines from the American Society of Transplantation Infectious Diseases Community of Practice. Clinical Transplantation. 2019;**33**(9):e13529

[52] De With K, Allerberger F, Amann S, Apfalter P, Brodt HR, Eckmanns T, et al. Strategies to enhance rational use of antibiotics in hospital: A guideline by the German Society for Infectious Diseases. Infection. 2016;**44**:395-439

[53] Lim CJ, Kong DC, Stuart RL. Reducing inappropriate antibiotic prescribing in the residential care setting: Current perspectives. Clinical Interventions in Aging. 2014;**9**:165-177

[54] Slama TG, Amin A, Brunton SA, File TM Jr, Milkovich G, Rodvold KA, et al. A clinician's guide to the appropriate and accurate use of antibiotics: The Council for Appropriate and Rational Antibiotic Therapy (CARAT) criteria. The American Journal of Medicine. 2005;**118**(7):1-6

[55] MacDougall C, Polk RE. Antimicrobial stewardship programs in health care systems. Clinical Microbiology Reviews. 2005;**18**(4):638-656

[56] Wong D, Spellberg B. Leveraging antimicrobial stewardship into improving rates of carbapenem-resistant Enterobacteriaceae. Virulence. 2017;**8**(4):383-390

[57] Rahman MM, Alam Tumpa MA, Zehravi M, Sarker MT, Yamin MD, Islam MR, et al. An overview of antimicrobial stewardship optimization: the use of antibiotics in humans and animals to prevent resistance. Antibiotics. 2022;**11**(5):667

[58] Sarwar MR, Saqib A, Iftikhar S, Sadiq T. Knowledge of community

pharmacists about antibiotics, and their perceptions and practices regarding antimicrobial stewardship: A cross-sectional study in Punjab, Pakistan. Infection and Drug Resistance. 2018;**11**:133-145

[59] Foral PA, Anthone JM, Destache CJ, Vivekanandan R, Preheim LC, Gorby GL, et al. Education and communication in an interprofessional antimicrobial stewardship program. Journal of Osteopathic Medicine. 2016;**116**(9): 588-593

[60] Yau JW, Thor SM, Tsai D, Speare T, Rissel C. Antimicrobial stewardship in rural and remote primary health care: A narrative review. Antimicrobial Resistance & Infection Control. 2021;**10**:1-33

[61] West LM, Cordina M. Educational intervention to enhance adherence to short-term use of antibiotics. Research in Social and Administrative Pharmacy. 2019;**15**(2):193-201

[62] Pearson M, Chandler C. Knowing antimicrobial resistance in practice: A multi-country qualitative study with human and animal healthcare professionals. Global Health Action. 2019;**12**(Supp 1):1599560

[63] Váradi L, Luo JL, Hibbs DE, Perry JD, Anderson RJ, Orenga S, et al. Methods for the detection and identification of pathogenic bacteria: Past, present, and future. Chemical Society Reviews. 2017;**46**(16):4818-4832

[64] Rittmann B, Stevens MP. Clinical decision support systems and their role in antibiotic stewardship: A systematic review. Current Infectious Disease Reports. 2019;**21**:1-2

[65] Taheri Moghadam S, Sadoughi F, Velayati F, Ehsanzadeh SJ, Poursharif S. The effects of clinical decision support system for prescribing medication on patient outcomes and physician practice performance: A systematic review and meta-analysis. BMC Medical Informatics and Decision Making. 2021;**21**(1):1-26

[66] Righi E, Mutters NT, Guirao X, Del Toro MD, Eckmann C, Friedrich AW, et al. ESCMID/EUCIC clinical practice guidelines on perioperative antibiotic prophylaxis in patients colonized by multidrug-resistant Gram-negative bacteria before surgery. Clinical Microbiology and Infection. 2023;**29**(4):463-479

[67] Hawkey PM, Warren RE, Livermore DM, McNulty CA, Enoch DA, Otter JA, et al. Treatment of infections caused by multidrug-resistant Gram-negative bacteria: Report of the British society for antimicrobial chemotherapy/healthcare infection society/british infection association joint working party. Journal of Antimicrobial Chemotherapy. 2018;**73**(Suppl. 3):iii2-iii78

[68] Tacconelli E, Sifakis F, Harbarth S, Schrijver R, van Mourik M, Voss A, et al. Surveillance for control of antimicrobial resistance. The Lancet Infectious Diseases. 2018;**18**(3):e99-e106

[69] Morel CM, De Kraker ME, Harbarth S. Enhanced Surveillance Expert Consensus Group (CANSORT-SCI). Surveillance of resistance to new antibiotics in an era of limited treatment options. Frontiers in Medicine. 2021;**8**:652638

[70] World Health Organization. Global Antimicrobial Resistance Surveillance System (GLASS) Report: Early Implementation 2016-2017. WHO. 2018. ISBN: 9789241513449

[71] Perez F, Villegas MV. The role of surveillance systems in confronting the global crisis of antibiotic-resistant

bacteria. Current Opinion in Infectious Diseases. 2015;**28**(4):375

[72] Tiseo G, Brigante G, Giacobbe DR, Maraolo AE, Gona F, Falcone M, et al. Diagnosis and management of infections caused by multidrug-resistant bacteria: Guideline endorsed by the Italian Society of Infection and Tropical Diseases (SIMIT), the Italian Society of Anti-Infective Therapy (SITA), the Italian Group for Antimicrobial Stewardship (GISA), the Italian Association of Clinical Microbiologists (AMCLI) and the Italian Society of Microbiology (SIM). International Journal of Antimicrobial Agents. 2022;**60**(2):106611

[73] Paul M, Carrara E, Retamar P, Tängdén T, Bitterman R, Bonomo RA, et al. European Society of Clinical Microbiology and Infectious Diseases (ESCMID) guidelines for the treatment of infections caused by multidrug-resistant Gram-negative bacilli (endorsed by European society of intensive care medicine). Clinical Microbiology and Infection. 2022;**28**(4):521-547

[74] Patel SV, Vergnano S. The impact of paediatric antimicrobial stewardship programmes on patient outcomes. Current Opinion in Infectious Diseases. 2018;**31**(3):216-223

[75] Jamaluddin NA, Periyasamy P, Lau CL, Ponnampalavanar S, Lai PS, Ramli R, et al. Point prevalence survey of antimicrobial use in a Malaysian tertiary care university hospital. Antibiotics. 2021;**10**(5):531

Chapter 11

Dispatching Biocompatible Polymers toward Antimicrobial Applications

Ruogu Tang

Abstract

Antimicrobial materials have become an essential part of various fields. In the past decades, various types of antimicrobial materials were developed and practically used. Based on the feedback from the clinical usage and market, the biocompatible materials have been very welcomed due to less side effects. This chapter provides a small and general review of biocompatible polymer materials and their applications in antimicrobial fields. This chapter could be divided into several parts: starting from the background introduction of microbial threats, the first section discusses the demands of biocompatible polymers for antimicrobial applications, then, the following sections would describe the basic knowledge of biocompatible polymers, including the definition, advantages, and typical examples, the next section reviewed and discussed some approaches to apply biocompatible polymers into antimicrobial applications.

Keywords: antimicrobial, biocompatibility, biocompatible polymer natural environment, functional modification, functional group

1. Introduction

Microbial invasions and infections have been accompanying human beings for thousands of years, and have caused countless causalities [1, 2]. The battle against microbial threats had already begun before people acquired the understanding of microorganisms [3]. During this long-lasting fighting, though the concept was not initially established, people had invented and developed antimicrobial materials, which contain contents that have the functions of preventing microbial invasion and/or sterilizing the microbes [4, 5]. Up to date, various types of antimicrobial materials have been developed and many of them have been successfully applied to real usages [6–9].

One major concern regarding the usage of antimicrobial materials is the side effects [10, 11]. For example, it was reported that some antimicrobial materials might cause damage to human bodies while killing the microbes [11]. Also, some antimicrobial materials, once abandoned or released, could lead to environmental and ecological issues [12]. Therefore, it has become a consensus that antimicrobial materials, no matter how potent they are, should be less collateral damaging. Under this demand, biocompatible polymers-based antimicrobial materials (i.e., biocompatible antimicrobial polymers) have become more favorable among customers and markets, and this type of material has become a new area for research [12].

This chapter provides a mini-review of biocompatible antimicrobial polymers, including the description of typical polymer materials, their usage in antimicrobial fields, the current challenges, and future perspectives.

2. Understanding the biocompatible polymers

2.1 Definition of biocompatible polymers

Admittedly, the terminology of biocompatible polymers is not well defined yet, and researchers prefer to create their own descriptions in the publications. However, so far, some consensuses regarding the definition of biocompatible polymers have been built and recognized among different researchers, which at least include the following:

1. The term "biocompatible polymer" refers to the material (or substance) with a polymeric structure, which consists of repeated sub-units that bond together [13]. This kind of material could be either obtained from nature or artificially synthesized.

2. The components of a given biocompatible polymer should not be intrinsically harmful to the target living recipients, including the contact sites (if any), local areas, and whole systems.

3. A biocompatible polymer, once applied to the recipients, should not cause any undesirable side effects. The material is supposed to provide appropriate host responses. In most applications, this requirement refers to both polymeric structures and the produced/released functional moieties and side products.

4. If the material remains on the target recipients for a longstanding period, this product should always be non-harmful until it is finally removed or fully reacted. Particularly for polymers, if they are degraded (either by design or not), the depolymerized components should also perform in the appropriate responses to the recipients.

2.2 Advantages of biocompatible polymers

It is easy to recognize that biocompatible polymers are attractive, but much more than the common opinions, biocompatible polymers have various advantages in different aspects. The following briefly describes some significant points:

Availability: unlike traditional polymers obtained from the synthesis of monomers, most biocompatible polymers (in the form of raw materials) could be obtained or extracted from natural products. For example, cellulose was rich in cotton, flax, and straw [14]; chitosan and chitin could be extracted in arthropod animals [15]; alginate could be obtained as a major product of brown algae [16]. The easy accessibility of biocompatible polymers reduces the costs, especially exempting the processing of monomer preparations and polymerizations.

Low hazard: as mentioned above, biocompatible polymers could leave fewer side effects; this was considered one of the most significant advantages. Within the antimicrobial application, the biocompatible polymers could be comfortably applied on humans without considering the toxicities or body exclusions, and there will be fewer concerns about byproducts during usage.

Biodegradability: for most traditional polymers, one critical shortage is collateral environmental pollution, as they are hard to decompose in the natural environment. However, many biocompatible polymers are biodegradable to microorganisms or medium, simplifying depolymerization's artificial processing after usage [17]. In addition, the degraded components are also generally biocompatible and could be further decomposed and absorbed by the microorganisms [17]. Therefore, the applications of biocompatible polymers could reduce environmental concerns.

2.3 Representative examples of biocompatible polymers

After years of development, certain biocompatible polymers have been commercially used, and more candidates are being investigated in the research labs. Among various products, they could be divided into natural and synthetic biocompatible polymers. This part listed some typical examples.

Cellulose: Cellulose is a linear polysaccharide that consists of 1,4-D-glucose units (**Figure 1**) [14]. It is one of the earliest used biocompatible polymers, even before people established the concept of polymer materials. Cellulose exists abundantly in nature and serves as a major component of the cell walls of plants and wood. As a natural product, cellulose is nontoxic, non-polluting, and fully biodegradable, making itself popular nowadays [18]. Within the antimicrobial applications, cellulose is more likely to be considered and used either as raw materials or substrate for further modification.

Chitin and chitosan: chitin and chitosan share a highly similar structure. Chitin is a linear polysaccharide that contains repeated N-acetyl-D-glucosamine units (**Figure 2**), and chitosan is a linear polysaccharide (**Figure 3**) composed of randomly distributed D-glucosamine (deacetylated unit) and N-acetyl-D-glucosamine (acetylated unit) [15]. Chitin and chitosan are also rich in nature, but unlike cellulose, chitin and chitosan are more likely to be found in arthropod animals such as crabs, shrimps, and some insects [15]. It is interesting to notice that chitin and some sorts of chitosan could be fully or partially digested by certain types of enzymes and decomposed into monosaccharides and ammonia [19]. The digestibility of chitosan and chitin has become a huge advantage for antimicrobial applications, as they have great potential for *in vivo* usage [19].

Alginic acid: alginic acid is a welcomed edible polysaccharide. Alginic acid has a copolymer structure of β-D-mannuronate block and α-L-guluronate block (**Figure 4**). In the natural environment, it often exists in the form of sodium salt (referred to as

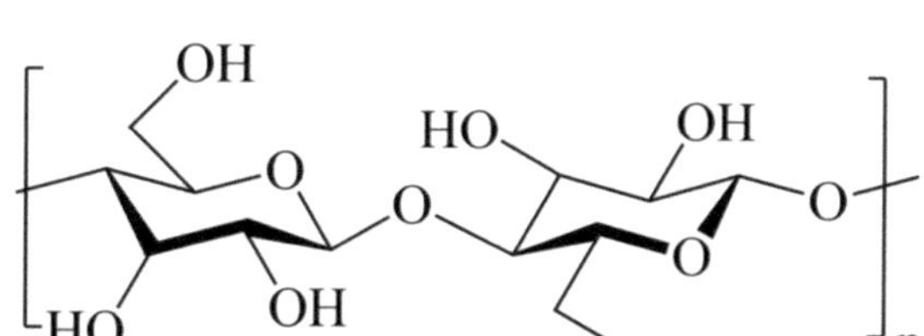

Figure 1.
Structure of cellulose.

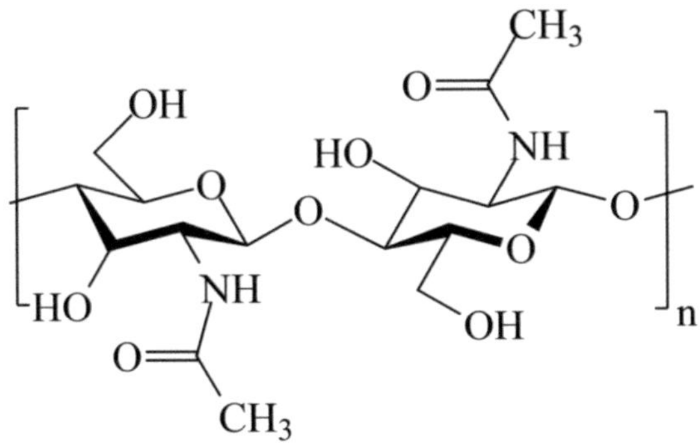

Figure 2.
Structure of chitin.

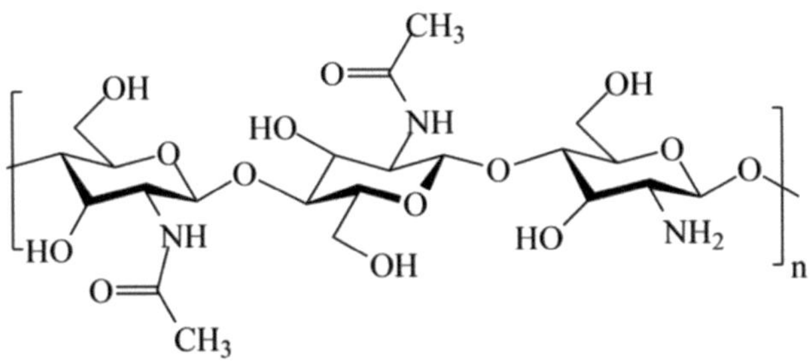

Figure 3.
Structure of chitosan.

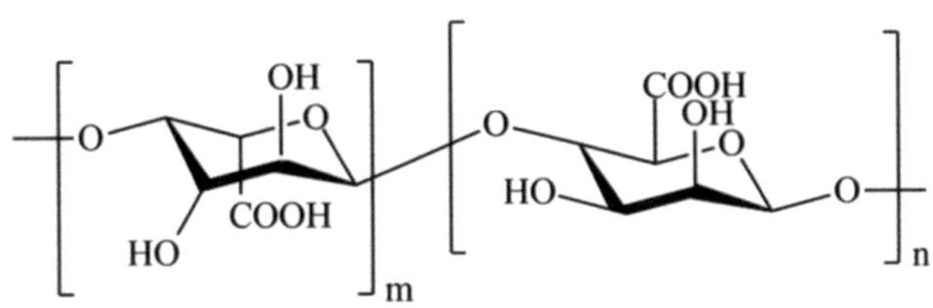

Figure 4.
Structure of alginic acid.

alginate) [16]. Thanks to its edibility, alginic acid is favorable in the food and drug industry and is becoming promising for antimicrobial applications.

Polyvinyl alcohol: polyvinyl alcohol (PVA) is a synthetic biocompatible polymer with the formula of $[CH_2CH(OH)]_n$. Though not invented as a biocompatible material, its biocompatibility was soon proved, and then, it started being investigated in biomedical engineering fields [20]. Unlike most polymers, PVA is water soluble, which could be fully carried forward in the case of antimicrobial application in aqueous conditions.

Of course, there are many more candidates than those listed above. To date, there is an increasing trend of research addressing biocompatible polymers [21], including obtaining novel materials, modifying current products, developing processing techniques, and exploring the markets.

3. Practicing biocompatible polymers in antimicrobial applications

3.1 Approaches toward antimicrobial applications

The intrinsic advantages of biocompatible polymers do not guarantee the feasibility of applications. Instead, the applications are largely associated with proper approaches. The following are some typical approaches developed for antimicrobial applications.

Direct processing: In terms of some biocompatible polymers, the raw materials could be extracted from natural products, and the raw materials could be directly manufactured to produce designed products without losing structure integrity, functionality, and biocompatibility. For example, chitosan with a high content of amino groups could be manufactured into thin films with antimicrobial functions due to the positively charged amino groups [22]. However, most biocompatible polymers in the form of raw materials do not have antimicrobial functions. Therefore, other approaches are needed.

Functional modification: For most biocompatible polymers, functional modification is the best approach to equip materials with antimicrobial abilities. A traditional and simple method is to physically blend polymers with antimicrobial moieties to obtain the antimicrobial mixtures, on which polymers could serve as the substrate materials to enhance the antimicrobial effects [23]. However, recent research has been more focused on the modifications of the polymer's own structures, as these kinds of materials are more chemically stable. Notably, many biocompatible polymers have functional groups on their backbones or side chains. These groups are chemically reactive and could be transferred into an antimicrobial moiety. For example, alginic acid and cyclodextrin could be oxidized, and the hydroxyl groups were transferred into aldehyde groups, which present some practical antimicrobial effects [24, 25]. Another method is to use crosslinkers, such as chitosan or other amino-containing polymers; the amino groups could be linked with crosslinkers containing aldehyde groups [26], acyl chloride [27], or other groups so the antimicrobial moieties could be covalently bound onto the polymers.

Transformation: It was found that materials in different forms would present different characteristics. This clue could be useful for biocompatible polymers. Many researches indicated that the raw polymers could be made into some "new" forms, such as nanoparticles [28], hydrogels [29], and colloids [30]. With a different size scale, shape, and morphology, the material might present antimicrobial functions or become a good candidate for antimicrobial modifications.

3.2 Representative antimicrobial applications of biocompatible polymers

A general recognition of biocompatible polymers for antimicrobial applications is the low and controllable side effects on targeted recipients while eliminating microorganisms. But when it comes to different specific fields, biocompatible polymers could present unique advantages in various ways; the following are some leading examples.

Medical and clinical usages: medical products made of biocompatible polymers have been top-rated in the market. Biocompatible polymer-based products, such as bandages, blood bags, and gauze, could be smoothly applied to humans. While eliminating and inhibiting microorganisms, the products could cause less toxic or hazardous chemicals and leave less hurtful or uncomfortable sensing. There would be fewer concerns about the residues from the products after usage.

Food industry: traditional polymer materials have always been used in the food industry (such as food containers, food packages, and food processing instruments), but in recent years, biocompatible polymers have become more favored. One major advantage of biocompatible polymers is less generating and releasing poisons onto food, enhancing food safety [31]. In addition, biocompatible polymers with antimicrobial functions could protect the food from microbe-induced spoilage and improve the food quality [32].

Environment engineering: in this field, biocompatible polymers were adequately used to produce biocides while killing microorganisms; they would not cause pollution or collateral damage to the surrounding environment. In addition, biodegradable products could be processed by natural decomposers, leaving fewer problems in the after-usage stage [33].

4. Conclusion

As a novel class of polymeric materials, biocompatible polymers have various unique and irreplaceable advantages for antimicrobial applications. However, the basic information of biocompatible polymers was not well recognized yet. This chapter provided a brief summary about biocompatible polymers and usages in antimicrobial applications, including a comprehensible and understandable definition, a list of advantages and some typical examples. Based on these, the achieved and potential approaches for antimicrobial applications were discussed. This chapter could serve as a “getting started” brochure to introduce the concept of using biocompatible polymers for antimicrobial applications to the public.

Conflict of interest

The author declared no conflict of interest.

Author details

Ruogu Tang
University of Massachusetts, Lowell, Massachusetts, USA

*Address all correspondence to: ruogutang@gmail.com

References

[1] Duan S, Wu R, Xiong YH, Ren HM, Lei C, Zhao Y, et al. Multifunctional antimicrobial materials: From rational design to biomedical applications. Progress in Materials Science. 2022;**125**:100887. DOI: 10.1016/j.pmatsci.2021.100887

[2] Yudaev P, Chuev V, Klyukin B, Kuskov A, Mezhuev Y, Chistyakov E. Polymeric dental nanomaterials: Antimicrobial action. Polymers. 2022;**14**(5):864. DOI: 10.3390/polym14050864

[3] Jurenas D, Fraikin N, Goormaghtigh F, Melderen LV. Biology and evolution of bacterial toxin–antitoxin systems. Nature Reviews Microbiology. 2022;**20**:335-350. DOI: 10.1038/s41579-021-00661-1

[4] Lanckohr C, Bracht H. Antimicrobial stewardship. Current Opinion in Critical Care. 2022;**28**(5):551-556. DOI: 10.1097/MCC.0000000000000967

[5] Siachalinga L, Mufwambi W, Lee IH. Impact of antimicrobial stewardship interventions to improve antibiotic prescribing for hospital inpatients in Africa: A systematic review and meta-analysis. Journal of Hospital Infection. 2022;**129**:124-143. DOI: 10.1016/j.jhin.2022.07.031

[6] Ahmed S, Sameen DE, Lu R, Li R, Dai J, Qin W, et al. Research progress on antimicrobial materials for food packaging. Critical Reviews in Food Science and Nutrition. 2022;**62**(11):3088-3102. DOI: 10.1080/10408398.2020.1863327

[7] Yang F, Wang A. Recent researches on antimicrobial nanocomposite and hybrid materials based on sepiolite and palygorskite. Applied Clay Science. 2022;**219**:106454. DOI: 10.1016/j.clay.2022.106454

[8] Sheridan M, Winters C, Zamboni F, Collins MN. Biomaterials: Antimicrobial surfaces in biomedical engineering and healthcare. Current Opinion in Biomedical Engineering. 2022;**22**:100373. DOI: 10.1016/j.cobme.2022.100373

[9] Haktaniyan M, Bradley M. Polymers showing intrinsic antimicrobial activity. Chemical Society Reviews. 2022;**51**:8584-8611. DOI: 10.1039/D2CS00558A

[10] Coombs K, Rodriguez-Quijada C, Clevenger J, Sauer-Budge A. Current understanding of potential linkages between biocide tolerance and antibiotic cross-resistance. Microorganisms. 2023;**11**(8):2000. DOI: 10.3390/microorganisms11082000

[11] Arzani F, dos Santos JHZ. Biocides and techniques for their encapsulation: A review. Soft Matter. 2022;**18**:5340-5358. DOI: 10.1039/D1SM01114F

[12] Weber F, Esmaeili N. Marine biofouling and the role of biocidal coatings in balancing environmental impacts. The Journal of Bioadhesion and Biofilm Research. 2023;**39**:661-681. DOI: 10.1080/08927014.2023.2246906

[13] Baranwal J, Barse B, Fais DGL, Kumar A. Biopolymer: A sustainable material for food and medical applications. Polymers. 2022;**4**(5):983. DOI: 10.3390/polym14050983

[14] Etale A, Onyianta AJ, Tuner SR, Eichhorn SJ. Cellulose: A review of water interactions, applications in composites, and water treatment. Chemical Reviews. 2023;**123**(5):2016-2048. DOI: 10.1021/acs.chemrev.2c00477

[15] Riseh RS, Vatankhah M, Hassanisaadi M, Kennedy JF. Chitosan-based nanocomposites as coatings and

packaging materials for the postharvest improvement of agricultural product: A review. Carbohydrate Polymers. 2023;**309**:120666. DOI: 10.1016/j.carbpol.2023.120666

[16] Saji S, Hebden A, Goswami P, Du C. A brief review on the development of alginate extraction orocess and its sustainability. Sustainability. 2022;**14**(9):5181. DOI: 10.3390/su14095181

[17] Samir A, Ashour FH, Hakim AAA, Bassyouni M. Recent advances in biodegradable polymers for sustainable applications. npj Materials Degradation. 2022;**6**:68. DOI: 10.1038/s41529-022-00277-7

[18] Tanpichai S, Boonmahitthisud A, Soykeabkaew N, Ongthip L. Review of the recent developments in all-cellulose nanocomposites: Properties and applications. Carbohydrate Polymers. 2022;**286**:119192. DOI: 10.1016/j.carbpol.2022.119192

[19] Zhou X, Zhang W, Zhao L, Gao S, Liu T, Yu D. Immobilization of lipase in chitosan-mesoporous silica material and pore size adjustment. International Journal of Biological Macromolecules. 2023;**235**:123789. DOI: 10.1016/j.ijbiomac.2023.123789

[20] Liu B, Zhang J, Guo H. Research progress of polyvinyl alcohol water-resistant film materials. Membranes. 2022;**12**(3):347. DOI: 10.3390/membranes12030347

[21] Das A, Ringu T, Ghosh S, Pramanik N. A comprehensive review on recent advances in preparation, physicochemical characterization, and bioengineering applications of biopolymers. Polymer Bulletin. 2022;**80**:7247-7312. DOI: 10.1007/s00289-022-04443-4

[22] Li D, Liu P, Hao F, Lv Y, Xiong W, Yan C, et al. Preparation and application of silver/chitosan-sepiolite materials with antimicrobial activities and low cytotoxicity. International Journal of Biological Macromolecules. 2022;**210**:337-349. DOI: 10.1016/j.ijbiomac.2022.05.015

[23] Lemke P, Junemann L, Moerschbacher BM. Synergistic antimicrobial activities of chitosan mixtures and chitosan-copper combinations. International Journal of Molecular Sciences. 2022;**23**(6):3345. DOI: 10.3390/ijms23063345

[24] Kusmono HP, Wildan MW. Oxidized cellulose nanocrystals from durian peel waste by ammonium persulfate oxidation. ACS Omega. 2023;**8**(33):30262-30272. DOI: 10.1021/acsomega.3c03117

[25] Varaprasad K, Karthikeyan C, Yallapu MM, Sadiku R. The significance of biomacromolecule alginate for the 3D printing of hydrogels for biomedical applications. International Journal of Biological Macromolecules. 2022;**212**:561-578. DOI: 10.1016/j.ijbiomac.2022.05.157

[26] Pawariya V, De S, Dutta J. Chitosan-based Schiff bases: Promising materials for biomedical and industrial applications. Carbohydrate Polymers. 2024;**323**:121395. DOI: 10.1016/j.carbpol.2023.121395

[27] Li B, Han L, Ma J, Zhao M, Yang B, Xu M, et al. Synthesis of acylated derivatives of chitosan oligosaccharide and evaluation of their potential antifungal agents on *Fusarium oxysporum*. Carbohydrate Polymers. 2023;**314**:120955. DOI: 10.1016/j.carbpol.2023.120955

[28] Ahmady AR, Solouk A, Saber-Samandari SS, Akbari S,

Ghanbari H, Brycki B. Capsaicin-loaded alginate nanoparticles embedded polycaprolactone-chitosan nanofibers as a controlled drug delivery nanoplatform for anticancer activity. Journal of Colloid and Interface Science. 2023;**638**:616-628. DOI: 10.1016/j.jcis.2023.01.139

[29] Liu W, Madry H, Cucchiarini. Application of alginate hydrogels for next-generation articular cartilage regeneration. International Journal of Biological Macromolecules. 2022;**23**(3):1147. DOI: 10.3390/ijms23031147

[30] Peers S, Montembault A, Ladaviere C. Chitosan hydrogels incorporating colloids for sustained drug delivery. Carbohydrate Polymers. 2022;**275**:118689. DOI: 10.1016/j.carbpol.2021.118689

[31] Zhang M, Biesold G, Choi W, Yu J, Deng Y, Silvestre C, et al. Recent advances in polymers and polymer composites for food packaging. Materials Today. 2022;**5**:134-161. DOI: 10.1016/j.mattod.2022.01.022

[32] Min T, Zhou L, Sun X, Du H, Zhu, Z, Wen Y. Electrospun functional polymeric nanofibers for active food packaging: A review. Food Chemistry. 2022;**391**:133239. DOI: 10.1016/j.foodchem.2022.133239

[33] Dalton B, Bhagabati P, De Micco J, Padamati RB, O'Connor K. A review on biological synthesis of the biodegradable polymers polyhydroxyalkanoates and the development of multiple applications. Catalysts. 2022;**12**(3):319. DOI: 10.3390/catal12030319

Chapter 12

Antimicrobial Resistance and Virulence of *Escherichia coli* in the Purview of Public Health Monitoring

Pankti Dhumal, Srashti Bajpai, Nachiket Garge, Agrima Bhatt, Fatema Rampurwala, Nishat Sulaimani, Shikha Gaikwad, Utpal Roy, Manasi Mishra and Rehan Deshmukh

Abstract

Antimicrobial resistance (AMR) has emerged as a major threat to human, animal, and environment health in the developed as well as the developing nations. The usage of antibiotics outside of the prescribed parameters in both the healthcare and livestock sectors is directly tied to this resistance event. Additionally, several *Escherichia coli* strains harbor the AMR genes, which can be transferred to humans leading to public health problems. Depending on the type of antibiotics used, *E. coli* has evolved to prowess several resistance mechanisms. Resistance genes that are horizontally transmissible also encode this resistance mechanism. Different resistance genes for each class of antibiotics are encoded by resistant *E. coli*. In conclusion, the current chapter ushers light on the molecular evolution of resistance and the regulatory genes contributing to the development of MDR in *E. coli*. Moreover, we have also discussed about the inappropriate practices of prescribing the antibiotics leading to intensifying the MDR in bacteria envisaging the implementation of rigorous guidelines for proper use of antibiotics in human beings.

Keywords: *E. coli*, antimicrobial resistance, genes, pathogenicity, virulence

1. Introduction

The primary concern that impacts public health in the twenty-first century is the resistance of pathogenic microorganisms to antibiotics [1]. This resistance is posing a more significant threat to public health worldwide [2]. Numerous classes of antibiotics can simultaneously render microorganisms resistant. The usage of antibiotics outside the prescribed parameters is linked to the emergence of antimicrobial resistance (AMR) [3]. The primary reasons fueling the egression of AMR in bacteria are its inappropriate use in agriculture, retail sectors like pharmacies, industries, and inefficacious prevention and control of infections in health care systems [4, 5].

The gastrointestinal tract of humans and animals is a reservoir of *Escherichia coli* which is a facultative anaerobe and usually, a harmless organism [6]. However, a few strains of *E. coli* have been known to cause infections in the gastrointestinal, urinary, and central nervous systems [7, 8]. Continuous exposure to antibiotics has shown to confer AMR in *E. coli* [9, 10]. Extended-spectrum β-lactamases (ESBLs) are plasmid-mediated enzymes that hydrolyze β-lactam antibiotics, including penicillin, cephalosporin, and the monobactam aztreonam, resulting in multidrug-resistant (MDR) organisms. Bacteria exhibiting ESBLs have been linked with unsatisfactory treatment outcomes [11]. MDR *E. coli* produces ESBLs is an example of antibiotic resistance linked to the infections that can be fatal [12]. Such AMR strains of *E. coli* present in animals are significant in developing infections in humans [6, 13–19]. Studies showed that through the consumption of contaminated food and water, AMR strains of *E. coli* can be transmitted from the environment to humans [9]. Therefore, evaluating the widespread of such MDR *E. coli* in various habitats becomes very necessary for setting guidelines in animal and human health care domains [17–19].

To this end, we conducted a systematic review and meta-analysis to investigate the prevalence, molecular evolution and the regulatory genes contributing to AMR in *E. coli*. We have also discussed the factors contributing to the development of AMR and possible ways to tackle and mitigate multidrug resistant in *E. coli*.

2. An overview of *E. coli*

E. coli is a Gram-negative facultative anaerobe which belongs to *Enterobacteriaceae* family of the Gammaproteobacteria class [20]. It is commonly found as normal microflora in human and animal gut [21]. Sequence analysis of the *E. coli* genome was first reported in 1997. Since then, more than 4800 *E. coli* genomes have been sequenced [22, 23]. *E. coli* has been widely used to monitor AMR in livestock and food of animal origin. This is because *E. coli* can be found in the digestive tracts of warm-blooded animals [24]. Furthermore, various strains of *E. coli* are probable sources of the AMR gene, which can be transmitted to humans through various means [21, 25]. *E. coli* carried through feces or treatment of wastewater disposed of in waterways can pollute the environment [26]. The concentration of *E. coli* per gram of feces differs over the host species, usually reaching 10^7–10^9 in humans and 10^4–10^6 in domestic animals [27]. Shiga toxin-producing *E. coli* (STEC), entero-toxigenic *E. coli* (ETEC), enteropathogenic *E. coli* (EPEC), diffusely adherent *E. coli*, and entero-invasive *E. coli* are among the six intestinal pathotypes of *E. coli*. These bacteria are categorized according to their pathogenicity mechanisms and virulence traits [28–31]. *E. coli* is a major cause of enteric foodborne illness, bloodstream infections, intra-abdominal infections, and urinary tract infections (UTIs) in animals and humans [30, 31]. UTIs can also be caused by *Proteus* spp., *Staphylococcus saprophyticus*, *Klebsiella* spp., and other *Enterobacteriaceae* in addition to *E. coli*. However, *E. coli* is regarded as the most prevalent source of nosocomial and community-acquired UTI among bacteria that cause UTIs. Additionally, patients' history of UTI raises their risk of infection [22]. Antibiotics such as cotrimoxazole (trimethoprim/sulfamethoxazole), nitrofurantoin, ciprofloxacin, and ampicillin are used as therapeutic methods for treating UTIs [32, 33]. However, research on antibiotic resistance in *E. coli* isolates from the urinary tract have revealed a rise in resistance to several antibiotics, such as ampicillin and cotrimoxazole [32]. Bloodstream infections brought on by *E. coli* can result in morbidity, death, and other health issues [34]. These infections may raise hospital mortality rates and result in antibiotic resistance, which lengthens hospital stays [35]. There have been reports of nosocomial bloodstream infections brought on by resistant bacteria, particularly in fragile and high-risk environments like critical care units and pediatric departments [36–38].

3. Molecular evolution of AMR in *E. coli*

AMR has been found to stem from human activities and indiscriminate or irrational uses of antibiotics. Various studies at genomic level pertaining to bacterial commensals and environmental bacteria revealed the presence of significant numbers of resistance determinants already embedded within their genomes not generated from any horizontal transmission and prior to the clinical introduction of antibiotics [39]. The molecular mechanism of emergence of antibiotic resistance and its spread, and persistence in bacterial population may be influenced by the interplay of several factors, such as (*a*) the mutation supply rate, (*b*) the level of resistance conferred by the resistance mechanism, (*c*) the fitness of the antibiotic-resistant mutant bacteria as a function of drug concentration, and (*d*) the strength of selective pressures [40]. Mutation supply rate is said to be determined by population sizes and mutation rates and horizontal gene transfer (HGT). Besides, in human hosts, the extent of genetic heterogeneity in a bacterial population is largely influenced by the mutation supply rate and the host-pathogen interaction [41]. The genetic perspective of AMR remains incomplete and impedes progress towards better patient outcomes and new therapeutics for resistant bacteria [42]. AMR surveillance drive indicated that drug-resistance to all the important antibiotics is in circulation among *E. coli* strains [43], including extended-spectrum β-lactams (ESBL), carbapenems, and more recently, plasmid-mediated colistin resistance (mcr-1), particularly in food animals from Asian countries [44–46]. As inter- and intraspecies HGT and mobile genetic elements are factors for AMR [47], close genomic surveillance of AMR cargo within *E. coli* populations is need of the hour. The mobilization of cargo genes from a donor to a recipient site essentially provides a contribution to HGT [48]. Transposon7 (Tn7)-carrying cargo and as many as 50 Tn7-like transposons identified integrons with antibiotic resistance gene cassettes and heavy metal resistance genes [49].

Mutations are also responsible for the acquisition of amino acid substitutions leading to change of properties of protein structures to provide an altered function and a selective thriving advantage [50]. The newer introduction of more potent antibiotics into clinical settings could potentially trigger the evolution of antibiotic-ring hydrolyzing enzymes rendering powerful drug resistance. These enzymes achieve the ability towards drastically decimating the prowess of antibiotics by acquiring point mutations near or at their active sites that augment catalytic activity of β-lactamases [51]. Since enzymes are endowed with a limited capacity of taking fewer destabilizing mutations before unfolding or loosing activity, newer gain-of-function mutations at active sites often comes at the cost of stability of the enzymes [52, 53]. The alanine to valine (A77V) substitution is such a stabilizing (point) mutation that demonstrates the compensation for a loss in stability associated with substitutions that alter catalytic prowess. This compensation mechanism enables the enzymes to continuum of evolution, resulting in elevated resistance [54].

The unresolved teething problem here is to predict the probability of HGT leading to antibiotic resistance. In this regard, various related factors might play the roles in influencing the probability where particular gene would transfer into a relevant human pathogen have not been very well understood [55]. The global microbiome might act as a potential source of resistance genes and genes for most classes of antibiotics have been found out in the human gut microbiome [56], and the soil microbiome [57].

Quiet interestingly, it has been hypothesized that within the human microbiota, there has been frequent exchanges between bacteria of antibiotic resistance genes (ARGs), where the intestinal *E. coli* community acts as a melting pot for the HGT [58, 59]. Antibiotic-sensitive bacterial populations developed drug-resistance either through HGTs or mutations and expression of resistance genes from other strains, either distantly or closely related (**Figure 1**). Detection of multiple mutational events can be selected in a step wisely to "train" resistance in a bacterium with successive mutations imparting additive effects [60]. As mentioned before, the human or animal gastrointestinal tract provides an ideal milieu where factors responsible for the upheaval of antibiotic resistance genes and spreading across resident bacterial populations are present [61]. The presence of high cell density is certainly one of the major factors. Additionally, spread of resistant infections is encouraged by selection followed by the innate ability for gene transfer through a variety of different mechanisms [62]. As bacterial cells get exposed to an anti-bacterial drug at sub minimum inhibitory concentrations (MIC), bacteria can thrive to gain resistance under selective pressure [63]. Recent work in this line demonstrated that there may be different resistance mechanisms induced by sub-inhibitory antibiotic exposure compared to lethal selection [64]. In *Salmonella enterica*, sub-inhibitory concentrations of streptomycin selected for high-level resistance through multiple-negligible-effect resistance

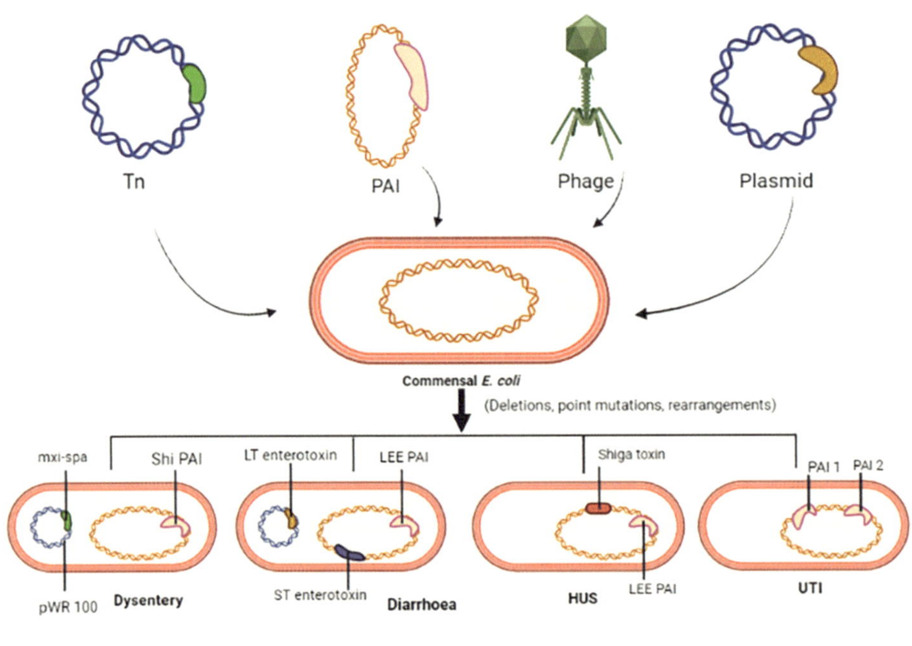

Figure 1.
E. coli virulence factors can be encoded by plasmids (e.g., heat-labile enterotoxin (LT) of ETEC and invasion factors of EIEC), several mobile genetic elements, including transposons (Tn) (e.g., heat stable enterotoxin (ST) of ETEC), pathogenicity islands (PAIs)- e.g., the locus of enterocyte effacement (LEE) of EPEC/EHEC and PAIs I and II of UPEC and bacteriophage (e.g., Shiga toxin of EHEC). Deletions, additions, and other genetic rearrangements may result in evolution of wild type to pathogenic E. coli strains responsible for causing dysentery (EIEC), diarrhea (EPEC, EAEC, EHEC, DAEC), urinary tract infections (UPEC) and hemolytic uremic syndrome (EHEC). UTI, urinary tract infection; HUS, hemolytic uremic syndrome. The figure was created using www.BioRender.com.

mutations, whereas lethal selection led to specific target mutations [65]. However, resistant mutants derived from sub-inhibitory fluoroquinolone exposure may not always induce causative changes in the target quinolone resistance determining region (QRDR) [65]. Experimental demonstrations also suggest that alterations in drug efflux systems are a major mechanism under pressure from sub-lethal ciprofloxacin exposure in case of *E. coli* [66]. Concentration-dependent responses are also mooted as anthropic drivers of resistance. These phenomena go beyond changes in resistance genes since different drug concentrations may alter the community behavior in response to drugs [67]. Reports are rife to indicate that for the manifestations of virulence and AMR, *E. coli* are armed with mobile genetic elements such as plasmids, genomic islands (PAIs and resistance islands [REIs]), or transposons [68]. The ability of mobile genetic elements is acquired at a higher rate [69] rendering *E. coli* ST131 an ideal subject to examine the co-evolution of AMR and virulence [70]. In such *E. coli* ST131 population, chromosomal mutations in quinolone-resistance determining regions (QRDR) of gyrA and parC, conferring high-level fluoroquinolone resistance were reported in the phylogenetic subclades, C1, and C2 [71, 72].

4. Regulatory genes in MDR *E. coli*

The emerging MDR of *E. coli* to several antibiotics is a significant problem leading to the development of pan resistant species and is the major bottleneck in the treatment of diseases. Development of resistance in bacterial pathogens is an adaptive trait and is acquired mainly either by HGT of mobile genetic elements carrying resistance or de novo mutations. *E. coli* strains producing ESBL enzymes like Cefotaximase-Munich (CTX-M) β-lactamases are emerging with higher frequency in hospital and healthcare-associated setups leading to community-onset urinary tract and bloodstream infections [73]. In many Gram-negative bacteria, chromosomal loci specifying global regulatory proteins that control MDR have been characterized in bacteria including *E. coli, Klebsiella pneumoniae,* and other *Enterobacteriaceae*, and *Neisseria gonorrhoeae*, *Bacillus subtilis,* and *S. aureus* [74]. In most instances, the regulatory proteins function as activators or repressors of transcription and contain either a single DNA-binding domain or two separate regions responsible for DNA binding and interacting with small-molecule substrates [75]. The potential genes responsible for encoding MDR in *E. coli* are listed in **Table 1**.

ESBLs hydrolyze and cause resistance to oxyimino-cephalosporins (e.g., cefotaxime, ceftazidime, ceftriaxone, cefuroxime and cefepime), monobactams (e.g., aztreonam), aminoglycosides and co-trimoxazole. ESBLs have evolved due to point mutations around the active site of β-lactamase [90]. Reports have described *ampC* genes present on the plasmid can be potentially transferred among *E. coli*, *Klebsiella* spp., and *Salmonella* spp. [76]. The ESBL/*ampC* genes are often surrounded by mobile genetic elements (e.g., transposons, IS elements or class 1 integrons), that mediate transmission, and involved in the expression of the genes IS*Ecp1* and IS*CR1* [77]. Gene encoding for CTX-M β-lactamases, $bla_{\text{CTX-M}}$ is encoded by narrow host-range incompatibility type plasmids (i.e.IncFI, IncFII, IncHI2 and IncI) or the broad host-range incompatibility type plasmids (i.e. IncN, IncP-1-a, IncL/M and IncA/C) and is associated with highly efficient mobile genetic element IS*Ecp1* [78–80]. Insertion sequence IS*Ecp1* belongs to the chromosomes of the environmental bacteria called *Kluyvera spp*. [91]. Plasmids encoding for CTX-M-15 also carry additional antibiotic resistance genes—$bla_{\text{OXA-1}}$, $bla_{\text{TEM-1}}$, *tetA*, *aac(6')-Ib-cr* and *aac(3)-II*, often

Sr. No	MDR Gene(s)	Localization of Gene	Function of Gene	Reference
1	*ampC*	Plasmid	Mediate resistance to cephalothin, cefazolin, cefoxitin, most penicillin, and beta-lactamase inhibitor-beta-lactam combinations.	[76]
2	*ISEcp1*	Plasmid	Mobilize adjacent DNA sequences by a so-called one-ended transposition mechanism.	[77]
3	*ISCR1*		Promoters in the expression of antibiotic resistance genes.	[77]
4	*blaCTX-M*	Plasmids	Promotes resistance to several antibiotics.	[78–80]
5	*IncFI*	Plasmid	Confers resistance to many antibiotic molecules.	[78–80]
6	*IncFII*	Plasmid	Confers resistance to many antibiotic molecules.	[78–80]
7	*IncHI2*	Plasmid	Responsible for mediating the horizontal transfer of numerous classes of resistance genes.	[78–80]
8	*IncI*	Plasmid	Mediating the horizontal transfer of numerous classes of resistance genes.	[78–80]
9	*IncP-1-a*	Plasmid	Known to spread genes between distinct phylogenetic groups of bacteria.	[78–80]
10	*IncL/M*	Plasmid	Role in the spread of genes encoding drug-resistance factors.	[78–80]
11	*ST-131*	Plasmid	Predominantly responsible for this global FQ-R and ceph-R pandemic causing millions of AMR infections annually.	[81]
12	*MCR-1*	Plasmid	Provides instructions for making a protein called the melanocortin 1 receptor.	[82]
13	*pp-flo*	Conjugative Plasmids	Facilitates their fast and efficient dissemination among bacteria of same or different species and genera.	[83]
14	*floSt*	Conjugative Plasmids	Facilitates their fast and efficient dissemination among bacteria of same or different species and genera.	[83]
15	*flo*	Conjugative Plasmids	Facilitates their fast and efficient dissemination among bacteria of same or different species and genera.	[83]
16	*floR*	Conjugative Plasmids	Facilitates their fast and efficient dissemination among bacteria of same or different species and genera.	[83]
17	*Mef(B)*	Transposable element	macrolide-efflux pump which mediates resistance to macrolides.	[84]
18	*blaOXA-1*	Contained within class 1 integrons	Hydrolyzes broad-spectrum cephalosporins.	[85]
19	*blaTEM-1*	Contained within class 1 integrons	Ampicillin resistance.	[85]

Sr. No	MDR Gene(s)	Localization of Gene	Function of Gene	Reference
20	*tetA*	Contained within class 1 integrons	Resistance to tetracycline by an active tetracycline efflux.	[85]
21	*aac(6′)-Ib-cr*	Contained within class 1 integrons	Conferring resistance to tobramycin, amikacin, and kanamycin.	[85]
22	*aac(3)-II*	Contained within class 1 integrons	Catalyze the covalent acetylation of a wide variety of aminoglycosides.	[85]
23	*cat I*	Chromosome	Inactivate chloramphenicol, thiamphenicol, azidamfenicol, and florfenicol.	[86]
24	*cat A*	Chromosome	provides instructions for making pieces (subunits) of catalase	[86]
25	*cat B*	Chromosome	provides instructions for making pieces (subunits) of catalase.	[86]
26	*etk*	Long arm of chromosome 7 at position 7q31.2	Regulating the polymerization and transport of virulence-determining capsular polysaccharide (CPS).	[87]
27	*AcrAB*	Upstream of *acrAB* operon	Preserve resistance acquisition by plasmid transfer in the presence of antibiotics with other modes of action.	[88]
28	*ISEcp1*	255 bp upstream of *blaCTX-M-15*	Mobilize adjacent DNA sequences by a so-called one-ended transposition mechanism.	[89]

Table 1.
List of genes encoding multiple drug resistance in E. coli.

contained within class 1 integrons [79, 85]. ESBL-producing Enterobacteriaceae are now significantly emerging in the community. The prototype *cat* genes (*cat I*, *cat II*, *cat III*, *cat A*, *cat B*, etc.) encode for chloramphenicol acetyltransferases (CAT) which can inactivate chloramphenicol as well as thiamphenicol, azidamfenicol, and florfenicol [86]. CATs are detected in a wide variety of bacteria including *E. coli*. *cat I* was originally identified as part of transposon Tn9 in *E. coli* and has been detected in several resistance plasmids of Gram-negative bacteria [92]. Various groups of *cat* genes are often located on small multicopy plasmids also carrying streptomycin resistance or macrolide resistance gene [93]. Export of chloramphenicol and florfenicol out of the bacterial cell via specific membrane transporters or multidrug transporters is another mechanism of resistance shown by microorganisms. Genes encoding for such transporters/efflux proteins (E-1 to E-8) are found in several clinically relevant and environmental bacteria [94]. A number of genes [*pp-flo*, *cmlA*-like, *floSt*, *flo*, or *floR*] grouped in E-3 mediate combined resistance to chloramphenicol and florfenicol. These genes show high homology [96–100%] and their protein products also show 88–100% identity. Often these genes coding for CATs and specific transporters are located on conjugative plasmids, gene cassettes and associated with transposon elements. This facilitates their fast and efficient dissemination among bacteria of same or different species and genera.

Sulphonamide resistance genes *sul1* and *sul2* encoding for dihydropteroate synthase enzymes have been known since decades [83]. A third sulphonamide resistance gene, *sul3*, was identified from porcine *E. coli* [95]. *Mef* (B) gene encoding for macrolide-efflux pump which mediates resistance to macrolides was found located in the vicinity of *sul3* in porcine *E. coli* [84]. In previous reports, mef (A), a homolog of mef (B) was found to be present only in Gram positive bacteria being encoded on a transposable element. However, the emergence of new resistance genes is *E. coli* isolates reflects on the positive selection pressure for sulphonamide and macrolide resistance despite its restricted use is many countries.

In *E. coli*, the MarA is the protein that regulates the expression of many chromosomal genes (the Mar regulon), including those specifying an MDR efflux pump and other proteins (e.g., porins) that mediate antibiotic susceptibility [96]. In *E. coli*, over expression of this MarA leads to reduced expression of the OmpF porin [97] as well as increased expression of the multidrug efflux pump AcrAB [88], thereby conferring resistance to a large number of antimicrobial agents. Resistance-nodulation-division (RND) pumps in *E. coli*, TolC plays the roles of proton antiporters and confer resistance to many important antibiotics, tetracyclines, chloramphenicol, some β-lactams, vancomycin, and fluoroquinolones [98, 99].

The most common ESBL gene in *E. coli* isolates of human origin is *blaCTX-M-15* and ST-131 clone and are implicated in AMR dissemination [81]. The increase in carbapenems (CPE) is mainly associated with the extensive dissemination of acquired CPE. CPE encoding genes are usually located in mobile genetic elements (MGEs), implying in the emergence of MDR and XDR strains [100]. Extraintestinal pathogenic *E. coli* (ExPEC) bacteria are the group of pathogenic strains that have the uncanny ability to cause severe ailments, such as urinary tract infections (UTIs) and bacteremia [101]. The emergence of MDR in *E. coli* has already become a global concern, with particular emphasis on *E. coli* sequence type (ST) 131 reported in UTIs. *CTX-M* Drug resistance phenomenon is mediated by extended-spectrum β-lactamases (ESBLs), mainly of the CTX-M family, particularly CTX-M-15 and 14, and less frequently of the SHV and OXA families [102]. CTX-M-15 first detected in *E. coli* isolates from India has become one of the most widely spread CTX-M β-lactamase worldwide. Association of $bla_{\text{CTX-M-15}}$ with the insertion element IS*Ecp1* plays an important role in its expression and mobilization [89]. From UTI patients, *E. coli* and *K. pneumoniae* with CTX-M type ESBLs were isolated. These enzymes are carried by *E. coli* with multidrug resistance phenotype [103]. MDR expressed by CTX-M-producing isolates from the community is often associated with the presence of multiple ESBLs genes, as well as aminoglycoside and quinolone resistance genes, thereby limiting the choice of effective antimicrobial drugs [104]. As mentioned earlier, that evolution in antibiotic resistance enzymes could rely on the acquisition of stabilizing mutations. A77V substitution is one such mutation that was detected in CTX-M extended-spectrum β-lactamases (ESBLs) from a good number of clinical isolates, pointing towards the significance of the β-lactamase evolution in antibiotic resistance in gram-negative bacteria [105].

Resistance to polypeptide antimicrobials like Colistin (CST) have also been reported in *E. coli* isolated from diseased pigs [106]. Most of such isolates also show resistance to sulfonamides, trimethoprim, tetracycline, ampicillin, or chloramphenicol. Acquired resistance to CST is known to be because of lipopolysaccharide (LPS) modifications—the addition of 2-aminoethanol, 4-amino-4-deoxy-L-arabinose (L-Ara4N) or phosphoethanolamine (PetN), or other strategies such as efflux pump and capsule formation. Mutations in chromosomal genes *mgrB*, *mgrR*, *etk* play a role in resistance to CST in *E. coli* [87]. Resistance to colistin could also be due to mutations in chromosomal genes or it may be acquired. *Colistin-resistance in E. coli* is plasmid-mediated colistin resistance via the acquisition of the *MCR-1* gene that can propagate efficiently and imparts resistance to other bacteria [107]. *MCR-1* protein enables the addition of a phosphor-ethanolamine group to lipid A. This modification imparts a charge alteration on LPS, which in turn reduces the binding affinity of colistin towards LPS [82].

5. AMR in *E. coli*: contributory factors and its mitigation

MDR *E. coli* spreads over time and stop responding to the drugs which makes it difficult to control the infections caused by them. This increases the risk of spreading diseases, developing major illnesses, and their transmission from one person to other. Hence, uttermost attention is required to mitigate MDR as it plays a significant role in the accomplishment of the sustainable development goals (SDGs). There are numerous factors (**Figure 2**) which contribute to AMR in *E. coli* listed below.

5.1 Overuse, misuse, and inappropriate prescribing of antibiotic or antimicrobials (AM)

It has been established that the main factors contributing to medication fatalities include taking antibiotics or AM specifically when not needed or ingesting it more than approved doses and concentrations. For example, any antibiotic used to treat a bacterial infection will kill susceptible bacteria; if appropriately targeted, the pathogenic microorganism will also be eradicated; nevertheless, the antibiotic will also

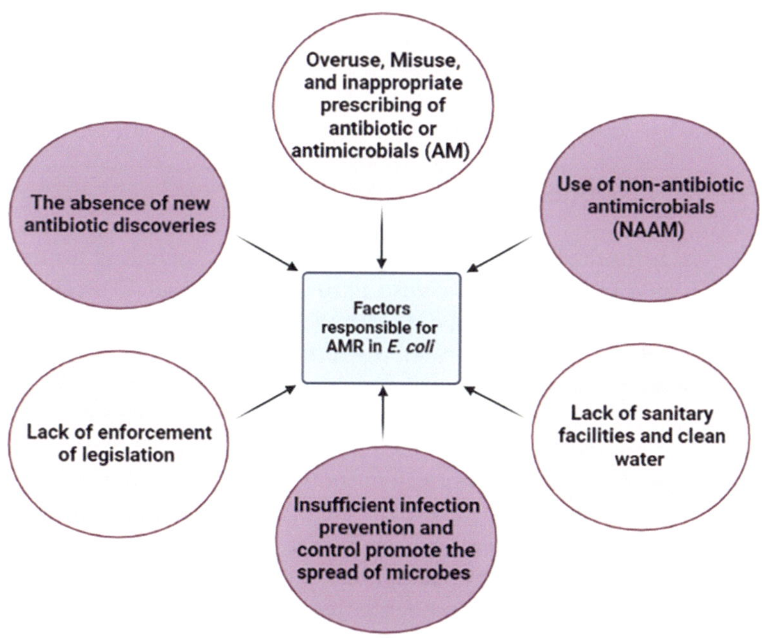

Figure 2.
Different factors responsible for developing the antimicrobial resistance in E coli.

kill any sensitive microbiota in the patient along with the infecting bacteria [108]. If the anatomical region has robust bacteria, whether they come from the healthy microbiota or the pathogenic germs that are being treated, they will proliferate and finally take control. As a result, bacteria are evolving quickly in response to selective antibiotic pressure, which is a significant feature in the growth of multidrug-resistant strains [108]. According to Balbina et al. [109], kanamycin, gentamicin, ciprofloxacin, fluoroquinolones, rifampicin, sulfisoxazole, cefoxitin, ampicillin, and fosfomycin are the most frequently prescribed antibiotics for treating *E. coli* related infections. However, ciprofloxacin is one of the drugs that are most likely to be prescribed incorrectly [109, 110]. Globally, the percentage of human antibiotic use was 65 percent between 2000 and 2015, however, if current trends are not changed, it is predicted that animal antibiotic use will increase by 11.5 percent by 2030, which will contribute to the emergence of additional MDR *E. coli* strains [111]. In the year 2021, Browne and his coworkers [112] reported that the global consumption rate of antibiotics was 14.3 well-defined daily doses (DDD) per thousand people per day in 2018 (40.2 [372–437] billion DDD), up 46% from 9.8 (92–105) DDD per 1000 per day in 2000 [112]. *E. coli* and other strains of the *Enterobacteriaceae* family are becoming gradually resistant to antibiotics due to an intensification in the appearance of membrane proteins that impel the drugs out of the cell, especially in topoisomerases, alterations in the target enzymes, and the presence of mobile genetic elements (plasmids, transposons, and integrons) that support in the lateral transmission of resistance genes in prokaryotes [113–115].

5.2 Use of non-antibiotic antimicrobials (NAAM)

Compared to antibiotics, a lot more non-antibiotic antimicrobial (NAAM) compounds are utilized as biocidal agents. As a result, the environment has significant residual levels of NAAM compounds in the soil, water, and air. One typical biocidal agent, triclosan (TCS), is utilized in more than 2000 different items, including hand washing liquids and toothpaste [116]. According to reports, triclosan causes *E. coli* to develop heritable multi-drug resistance [116]. 1.1×10^5 to 4.2×10^5 kg of TCS is released annually by wastewater treatment plants (WWTPs) in the US alone [117].

5.3 Lack of sanitary facilities and clean water

Water is a vital resource for all living things. All species, including humans, have microbes living inside and on them. According to Collignon et al. [118], environmental variable quantity such as advanced deficient technologies in waste management, availability of non-potable drinking water, overcrowded housing, and poor cleanliness may stimulate the growth and expansion of resistant bacteria globally. The unique capability of *E. coli* in colonization of the animal gut has given additional evolutionary benefit in acquiring resistance attributes from other bacteria in its environment [118]. Antibiotics are frequently used in livestock or food animals to combat and prevent diseases as well as to promote growth, which causes AMR to develop and raises the possibility of antibiotic-resistant bacterial colonization [118]. These bacteria may be spread while handling the animals, slaughtering them, or processing the meat. Additionally, fruits and vegetables can be infected with fecal material directly from the animals or through irrigation water that has been contaminated with animal or human waste [116, 118]. More than 1000 unique antibiotic-resistant genes have been identified in the human gut microbiota, and the transfer of these features between gut commensals is prevalent, according to one study done by Hu et al. [119]. In underdeveloped regions of the world, antibiotic-resistant strains of *E. coli* are detected at the time of birth, at the highest rates in human feces. Studies in China [114] have stated that ESBL genes in *E. coli* are transmitted by conjugation that enters in food chain directly or indirectly which facilitates explaining why ESBL-producing *E. coli* are so prevalent in feces around the world.

5.4 Insufficient infection prevention and control promote the spread of microbes

Since many sick individuals congregate in hospitals and healthcare centers and antibiotic consumption is prevalent, resistant strains of bacteria are selected for and disseminated throughout these settings, making them important hotspots for most bugs [120]. Additionally, inadequate hygiene habits may make it easier for resistant microbes to transmit from patients or visitors to doctors, nurses, and other healthcare professionals through their hands or clothing [121]. Other risk factors include inadequate sanitation, inappropriate cleaning of the facilities, and dirty equipment. Few isolation rooms and crowded wards further encourage spread [120, 121]. Poor approaches towards reasonably priced medicines, vaccines, and diagnostics are the most defined factors for showing the resistance to antimicrobial treatment case of *E. coli*, which is understudied in many underdeveloped nations [122]. Patients face substantial health risks when doctors attempt to treat infections about which they are ignorant. Medical facilities should set up systems to guarantee the identification and management of a range of bacterial diseases without prescribing incorrect medications [122]. Over time, incorrect medicine prescriptions have led to several patients' deaths and even more severe health issues [120–122] With the right diagnosis and care, patients can reduce their risk of developing medication resistance-related issues. In addition, the ongoing development of strategies for the cultivation of bacteria using a variety of culture media and conditions a process known as culturomics is crucial for the identification of antibiotic resistance genes (ARGs) and the comprehension of the cellular mechanisms underlying the association between ARGs and resistance phenotypes in various bacteria [123]. One of the major issues with building a consolidated database is establishing a clear strategy and making it into a regular practice of environmental-quality monitoring [124].

5.5 Lack of enforcement of legislation

By 2050, it is predicted that antibiotic resistance will have killed more people than cancer today. According to Naghavi's analysis in the published report in the year 2022, there would be 12.7 million deaths caused by bacterial AMR in 2022, out of an expected 495 million deaths linked to bacterial AMR in 2019 [125]. Death by third-generation cephalosporin-resistant *E. coli* was found to be the most prevalent [125]. Therefore, the lawmakers must keep an eye on how often antibiotics or other biocidal chemicals are used. The sum of all the efforts, including legislation, studies of community attitudes and knowledge regarding antibiotics, educational initiatives, and follow-up policies, will undoubtedly contribute not only to reducing antibiotic misuse in the short term but also to fostering appropriate knowledge among society members regarding the advantages of using antibiotics properly. These actions will be crucial to lowering patient demand for antibiotics and preventing detrimental effects on public health and the economy. Although the creation of stringent legislation like "Schedule H1" is crucial, it might only be one component of the solution. Delaying these reforms emphasizes the risky game we are playing with the health of people and animals, which might have terrible repercussions.

5.6 The absence of new antibiotic discoveries

It can be crucial in treating a variety of bacterial illnesses. Some bacterial strains are still evolving and developing medication resistance. Although few new antibiotics are being introduced to the market, they could be useful in attempting to close the gap.

6. Conclusions and future horizons

A persistent global concern in the field of medicine is microbial resistance to antibiotics. One among numerous bacteria is *E. coli*, which is frequently found in humans and animals. Finding novel antibiotic substitutes that are more potent in treating pathogenic infections, particularly the ones caused due to *E. coli* is required since there are less options for antibiotics that are sensitive to this bacterium. *E. coli* and other Gram-negative bacteria have evolved to develop different mechanism to encounter any antimicrobials. Moreover, different studies have showed higher prevalence of ESBL antibiotic resistance and MDR *E. coli* strains in animals than in humans. Besides, the stability and spread of colistin resistance in multiple reservoirs are an emerging public health threat worldwide. Dissemination of antibiotic resistance genes via transmissible plasmids, integrons and transposons to several environmental reservoirs like wastewater treatment plants (WWTPs), agriculture fields, animal manure, and hospital-associated setups is an emerging threat to environmental and public health safety. Our meta-analysis suggests that comprehensive surveilling of hospital-associated diseases, careful monitoring of hospital waste management procedures, supervising antibiotic use in animals, tracking and evaluating antibiotics susceptibility trends, and the development of trustworthy antibiotic strategies may facilitate more corrective actions for the inhibition and control of *E. coli* infections in various parts of the world and practical guidelines can be put in place for proper prescription of antibiotics in humans to ensure better public health monitoring.

Acknowledgements

RD would like to thank Dr. Vishwanath Karad MIT-World Peace University for providing Students' Innovation Seed Fund to PD, SB, NG, AB, FR and NS. SG and MM are grateful to the institute for providing the facilities and infrastructure.

Conflict of interest

The authors declare no conflict of interest. All authors have contributed to the chapter and agreed for its publication.

DOI: http://dx.doi.org/10.5772/intechopen.108299

Author details

Pankti Dhumal[1], Srashti Bajpai[1], Nachiket Garge[1], Agrima Bhatt[1], Fatema Rampurwala[1], Nishat Sulaimani[1], Shikha Gaikwad[1], Utpal Roy[2], Manasi Mishra[1] and Rehan Deshmukh[1]*

1 Faculty of Science, School of Biology, Dr. Vishwanath Karad MIT World Peace University, Pune, India

2 Department of Biological Sciences, Birla Institute of Technology and Science-K.K. Birla Goa Campus, Goa, India

*Address all correspondence to: rrehandeshmukh@gmail.com

References

[1] Irrgang A, Hammerl JA, Falgenhauer L, Guiral E, Schmoger S, Imirzalioglu C, et al. Diversity of CTX-M-1- producing *E. coli* from German food samples and genetic diversity of the blaCTX-M-1 region on IncI1 ST3 plasmids. Veterinary Microbiology. 2018;**221**:98-104

[2] World Health Organization. Antimicrobial Resistance: Global Report on Surveillance. Geneva, Switzerland: WHO.int; 2014. https://doi.org/9789241564748

[3] Ramírez-Castillo FY, Moreno-Flores AC, Avelar-González FJ, Márquez-Díaz F, Harel J, Guerrero-Barrera AL. An evaluation of multidrug-resistant *Escherichia coli* isolates in urinary tract infections from Aguascalientes, Mexico: Cross- sectional study. Annals of Clinical Microbiology and Antimicrobials. 2018;**17**(1):1-3

[4] Alonso CA, Zarazaga M, Ben Sallem R, Jouini A, Ben Slama K, Torres C. Antibiotic resistance in *Escherichia coli* in husbandry animals: The African perspective. Letters in Applied Microbiology. 2017;**64**(5):318-334

[5] Van Boeckel TP, Brower C, Gilbert M, Grenfell BT, Levin SA, Robinson TP, et al. Global trends in antimicrobial use in food animals. Proceedings of the National Academy of Sciences. 2015;**112**(18):5649-5654

[6] Agunos A, Léger D, Carson C. Review of antimicrobial therapy of selected bacterial diseases in broiler chickens in Canada. The Canadian Veterinary Journal. 2012;**53**(12):1289

[7] Landoni MF, Albarellos G. The use of antimicrobial agents in broiler chickens. The Veterinary Journal. 2015;**205**(1):21-27

[8] Page SW, Gautier P. Use of antimicrobial agents in livestock. Revue Scientifique et Technique-OIE. 2012;**31**(1):145

[9] Chen Y, Liu Z, Zhang Y, Zhang Z, Lei L, Xia Z. Increasing prevalence of ESBL-producing multidrug resistance *Escherichia coli* from diseased pets in Beijing, China from 2012 to 2017. Frontiers in Microbiology. 2019;**10**:2852

[10] Effendi MH, Cicilia R, Rahmahani J, Tyasningsih W. Public awareness for antimicrobial resistance from *Escherichia coli* isolated from beef sold on several wet market in Surabaya, Indonesia. Indian Journal of Public Health Research and Development. 2020;**11**(9):295-300

[11] Baron S, Jouy E, Larvor E, Eono F, Bougeard S, Kempf I. Impact of third-generation-cephalosporin administration in hatcheries on fecal *Escherichia coli* antimicrobial resistance in broilers and layers. Antimicrobial Agents and Chemotherapy. 2014;**58**(9):5428-5434

[12] Chantziaras I, Boyen F, Callens B, Dewulf J. Correlation between veterinary antimicrobial use and antimicrobial resistance in food-producing animals: A report on seven countries. Journal of Antimicrobial Chemotherapy. 2014;**69**(3):827-834

[13] Peralta-Sánchez JM, Martín-Platero AM, Ariza-Romero JJ, Rabelo-Ruiz M, Zurita-González MJ, Baños A, et al. Egg production in poultry farming is improved by probiotic bacteria. Frontiers in Microbiology. 2019;**10**:1042

[14] Nhung NT, Chansiripornchai N, Carrique-Mas JJ. Antimicrobial resistance in bacterial poultry pathogens:

A review. Frontiers in Veterinary Science. 2017;**4**:126

[15] Chah KF, Ugwu IC, Okpala A, Adamu KY, Alonso CA, Ceballos S, et al. Detection and molecular characterisation of extended-spectrum β- lactamase-producing enteric bacteria from pigs and chickens in Nsukka, Nigeria. Journal of Global Antimicrobial Resistance. 2018;**15**:36-40

[16] Falgenhauer L, Imirzalioglu C, Oppong K, Akenten CW, Hogan B, Krumkamp R, et al. Detection and characterization of ESBL-producing *Escherichia coli* from humans and poultry in Ghana. Frontiers in Microbiology. 2019;**9**:3358

[17] Martínez-Vázquez AV, Rivera-Sánchez G, Lira-Méndez K, Reyes-López MÁ, Bocanegra-García V. Prevalence, antimicrobial resistance and virulence genes of *Escherichia coli* isolated from retail meat in Tamaulipas, Mexico. Journal of Global Antimicrobial Resistance. 2018;**14**:266-272

[18] Rahman SU, Ahmad S, Khan I. Incidence of ESBL-producing-*Escherichia coli* in poultry farm environment and retail poultry meat. Pakistan Veterinary Journal. 2018;**39**:116-120

[19] Casella T, Nogueira MC, Saras E, Haenni M, Madec JY. High prevalence of ESBLs in retail chicken meat despite reduced use of antimicrobials in chicken production, France. International Journal of Food Microbiology. 2017;**257**:271-275

[20] Jang J, Hur HG, Sadowsky MJ, Byappanahalli MN, Yan T, Ishii S. Environmental *Escherichia coli*: Ecology and public health implications—A review. Journal of Applied Microbiology. 2017;**123**(3):570-581

[21] Kheiri R, Akhtari L. Antimicrobial resistance and integron gene cassette arrays in commensal *Escherichia coli* from human and animal sources in IRI. Gut Pathogens. 2016;**8**(1):1-0

[22] Jan NA, Meshram SU, Kulkarni A. Plasmid profile analysis of multidrug resistant *E. coli* isolated from UTI patients of Nagpur City, India. Romanian. Biotechnological Letters. 2009;**14**(5):4635-4640

[23] Tenaillon O, Barrick JE, Ribeck N, Deatherage DE, Blanchard JL, Dasgupta A, et al. Tempo and mode of genome evolution in a 50,000-generation experiment. Nature. 2016;**536**(7615): 165-170

[24] Nhung NT, Cuong NV, Thwaites G, Carrique-Mas J. Antimicrobial usage and antimicrobial resistance in animal production in Southeast Asia: A review. Antibiotics. 2016;**5**(4):37

[25] Overdevest I, Willemsen I, Rijnsburger M, Eustace A, Xu L, Hawkey P, et al. Extended-spectrum β-lactamase genes of *Escherichia coli* in chicken meat and humans, The Netherlands. Emerging Infectious Diseases. 2011;**17**(7):1216

[26] Berthe T, Ratajczak M, Clermont O, Denamur E, Petit F. Evidence for coexistence of distinct *Escherichia coli* populations in various aquatic environments and their survival in estuary water. Applied and Environmental Microbiology. 2013;**79**(15):4684-4693

[27] Tenaillon O, Skurnik D, Picard B, Denamur E. The population genetics of commensal *Escherichia coli*. Nature Reviews Microbiology. 2010;**8**(3): 207-217

[28] Issa AH, Almayah AA. New virulence factor of Normal Flora *E. coli*. Systematic Reviews in Pharmacy. 2020;**11**(2): 71-76

[29] Kai A, Konishi N, Obata H. Diarrheagenic *Escherichia coli*. Nihon rinsho. Japanese Journal of Clinical Medicine. 2010;**68**:203-207

[30] Kaper JB, Nataro JP, Mobley HLT. Pathogenic E. coli. Nature Reviews/ Microbiology. 2004;**2**:123-140

[31] Alanazi MQ, Alqahtani FY, Aleanizy FS. An evaluation of *E. coli* in urinary tract infection in emergency department at KAMC in Riyadh, Saudi Arabia: Retrospective study. Annals of Clinical Microbiology and Antimicrobials. 2018;**17**(1):1-7

[32] Foxman B. The epidemiology of urinary tract infection. Nature Reviews Urology. 2010;**7**(12):653-660

[33] Poolman JT, Wacker M. Extraintestinal pathogenic Escherichia coli, a common human pathogen: challenges for vaccine development and progress in the field. The Journal of Infectious Diseases. 1 Jan 2016;**213**(6):6-13

[34] Naylor NR, Pouwels KB, Hope R, Green N, Henderson KL, Knight GM, et al. The health and cost burden of antibiotic resistant and susceptible *Escherichia coli* bacteraemia in the English hospital setting: A national retrospective cohort study. PLoS One. 2019;**14**(9):e0221944

[35] Saied T, Elkholy A, Hafez SF, Basim H, Wasfy MO, El-Shoubary W, et al. Antimicrobial resistance in pathogens causing nosocomial bloodstream infections in university hospitals in Egypt. American Journal of Infection Control. 2011;**39**(9): e61-e65

[36] Tansarli GS, Poulikakos P, Kapaskelis A, Falagas ME. Proportion of extended- spectrum β-lactamase (ESBL)-producing isolates among Enterobacteriaceae in Africa: Evaluation of the evidence—Systematic review. Journal of Antimicrobial Chemotherapy. 2014;**69**(5):1177-1184

[37] Ma F, Xu S, Tang Z, Li Z, Zhang L. Use of antimicrobials in food animals and impact of transmission of antimicrobial resistance on humans. Biosafety and Health. 2021;**3**(01):32-38

[38] Petrillo F, Folliero V, Santella B, Franci G, Foglia F, Trotta MC, et al. Prevalence and antibiotic resistance patterns of ocular bacterial strains isolated from pediatric patients in University Hospital Of Campania "Luigi Vanvitelli," Naples, Italy. International Journal of Microbiology. 2020:6

[39] Cafini F, Romero VM, Morikawa K. Mechanisms of horizontal gene transfer. The Rise of Virulence and Antibiotic Resistance in Staphylococcus Aureus. 2017;**61**:61-79

[40] Dunny GA, Yuhasz MI, Ehrenfeld EL. Genetic and physiological analysis of conjugation in *Streptococcus faecalis*. Journal of Bacteriology. 1982;**151**(2):855-859

[41] Worby CJ, Lipsitch M, Hanage WP. Within-host bacterial diversity hinders accurate reconstruction of transmission networks from genomic distance data. PLoS Computational Biology. 2014;**10**(3):e1003549

[42] Munita JM, Arias CA. Mechanisms of antibiotic resistance. Microbiology Spectrum. 2016;**4**(2):4-2

[43] Paitan Y. Current trends in antimicrobial resistance of *Escherichia coli*. *Escherichia coli*, a Versatile Pathogen. 2018;**416**:181-211

[44] Zhao Q, Li Y, Tian Y, Shen Y, Wang S, Zhang Y. Clinical impact of colistin banning in food animal on mcr-1-Positive enterobacteriaceae in patients from Beijing, China, 2009-2019: A long-term longitudinal observational study. Frontiers in Microbiology. 2022;**13**:826624

[45] Newton-Foot M, Snyman Y, Maloba MR, Whitelaw AC. Plasmid-mediated mcr-1 colistin resistance in Escherichia coli and Klebsiella spp. clinical isolates from the Western Cape region of South Africa. Antimicrobial Resistance & Infection Control. Dec 2017;**6**(1):1-7

[46] Sun J, Zeng X, Li XP, Liao XP, Liu YH, Lin J. Plasmid-mediated colistin resistance in animals: Current status and future directions. Animal Health Research Reviews. 2017;**18**(2):136-152

[47] Partridge SR, Kwong SM, Firth N, Jensen SO. Mobile genetic elements associated with antimicrobial resistance. Clinical Microbiology Reviews. 2018;**31**(4):e00088-e00017

[48] Peters JE, Fricker AD, Kapili BJ, Petassi MT. Heteromeric transposase elements: Generators of genomic islands across diverse bacteria. Molecular Microbiology. 2014;**93**(6):1084-1092

[49] Parks AR, Peters JE. Transposon Tn 7 is widespread in diverse bacteria and forms genomic islands. Journal of Bacteriology. 2007;**189**(5):2170-2173

[50] Romero PA, Arnold FH. Exploring protein fitness landscapes by directed evolution. Nature Reviews Molecular Cell Biology. 2009;**10**(12):866-876

[51] Bush K. Proliferation and significance of clinically relevant β-lactamases. Annals of the New York Academy of Sciences. 2013;**1277**(1):84-90

[52] Bloom JD, Labthavikul ST, Otey CR, Arnold FH. Protein stability promotes evolvability. Proceedings of the National Academy of Sciences. 2006;**103**(15):5869-5874

[53] Orencia MC, Yoon JS, Ness JE, Stemmer WP, Stevens RC. Predicting the emergence of antibiotic resistance by directed evolution and structural analysis. Nature Structural Biology. 2001;**8**(3):238-242

[54] Patel MP, Fryszczyn BG, Palzkill T. Characterization of the global stabilizing substitution A77V and its role in the evolution of CTX-M β-lactamases. Antimicrobial Agents and Chemotherapy. 2015;**59**(11):6741-6748

[55] Hughes D, Andersson DI. Environmental and genetic modulation of the phenotypic expression of antibiotic resistance. FEMS Microbiology Reviews. 2017;**41**(3):374-391

[56] Sommer MO, Dantas G, Church GM. Functional characterization of the antibiotic resistance reservoir in the human microflora. Science. 2009;**325**(5944):1128-1131

[57] D'Costa VM, McGrann KM, Hughes DW, Wright GD. Sampling the antibiotic resistome. Science. 2006;**311**:374-377. Google Scholar There is no corresponding record for this reference

[58] Trobos M, Lester CH, Olsen JE, Frimodt-Møller N, Hammerum AM. Natural transfer of sulphonamide and ampicillin resistance between *Escherichia coli* residing in the human intestine. Journal of Antimicrobial Chemotherapy. 2009;**63**(1):80-86

[59] Porse A, Gumpert H, Kubicek-Sutherland JZ, Karami N, Adlerberth I, Wold AE, et al. Genome dynamics

of *Escherichia coli* during antibiotic treatment: Transfer, loss, and persistence of genetic elements in situ of the infant gut. Frontiers in Cellular and Infection Microbiology. 2017;7:126

[60] Alekshun MN, Levy SB. Molecular mechanisms of antibacterial multidrug resistance. Cell. 2007;**128**(6):1037-1050

[61] Huddleston JR. Horizontal gene transfer in the human gastrointestinal tract: Potential spread of antibiotic resistance genes. Infection and Drug Resistance. 2014;7:167

[62] Bennett PM. Plasmid encoded antibiotic resistance: Acquisition and transfer of antibiotic resistance genes in bacteria. British Journal of Pharmacology. 2008;**153**(S1):S347-S357

[63] Andersson DI, Hughes D. Microbiological effects of sublethal levels of antibiotics. Nature Reviews Microbiology. 2014;**12**(7):465-478

[64] Ramsay KA, McTavish SM, Wardell SJ, Lamont IL. The effects of sub-inhibitory antibiotic concentrations on *Pseudomonas aeruginosa*: Reduced susceptibility due to mutations. Frontiers in Microbiology. 2021;**20**(12):789550

[65] Wistrand-Yuen E, Knopp M, Hjort K, Koskiniemi S, Berg OG, Andersson DI. Evolution of high-level resistance during low-level antibiotic exposure. Nature Communications. 2018;**9**(1):1-2

[66] Li XZ, Plésiat P, Nikaido H. The challenge of efflux-mediated antibiotic resistance in gram-negative bacteria. Clinical Microbiology Reviews. 2015;**28**(2):337-418

[67] Wang Q, Sun FJ, Liu Y, Xiong LR, Xie LL, Xia PY. Enhancement of biofilm formation by subinhibitory concentrations of macrolides in icaADBC-positive and-negative clinical isolates of *Staphylococcus epidermidis*. Antimicrobial Agents and Chemotherapy. 2010;**54**(6):2707-2711

[68] Kaper JB, Nataro JP, Mobley HL. Pathogenic *Escherichia coli*. Nature Reviews Microbiology. 2004;**2**(2):123-140

[69] Decano AG, Downing T. An *Escherichia coli* ST131 pangenome atlas reveals population structure and evolution across 4,071 isolates. Scientific Reports. 2019;**9**(1):1-3

[70] Totsika M, Beatson SA, Sarkar S, Phan MD, Petty NK, Bachmann N, et al. Insights into a multidrug resistant Escherichia coli pathogen of the globally disseminated ST131 lineage: Genome analysis and virulence mechanisms. PLoS One. 2011;**6**(10):e26578

[71] Biggel M, Moons P, Nguyen MN, Goossens H, Van Puyvelde S. Convergence of virulence and antimicrobial resistance in increasingly prevalent *Escherichia coli* ST131 papGII+ sub lineages. Communications Biology. 2022;**5**(1):1-0

[72] Piddock LJ, Hall MC, Walters RN. Phenotypic characterization of quinolone-resistant mutants of Enterobacteriaceae selected from wild type, gyrA type and multiplyresistant (marA) type strains. Journal of Antimicrobial Chemotherapy. 1991;**28**(2):185-198

[73] Nicolas-Chanoine MH, Blanco J, Leflon-Guibout V, Demarty R, Alonso MP, Caniça MM, et al. Intercontinental emergence of *Escherichia coli* clone O25: H4-ST131 producing CTX-M-15. Journal of Antimicrobial Chemotherapy. 2008;**61**(2):273-281

[74] Grkovic S, Brown MH, Skurray RA. Regulation of bacterial drug export

systems. Microbiology and Molecular Biology Reviews. 2002;**66**(4):671-701

[75] Schumacher MA, Brennan RG. Structural mechanisms of multidrug recognition and regulation by bacterial multidrug transcription factors. Molecular Microbiology. 2002;**45**(4):885-893

[76] Jacoby GA, Munoz-Price LS. The new β-lactamases. New England Journal of Medicine. 2005;**352**(4):380-391

[77] Poirel L, Naas T, Nordmann P. Genetic support of extended-spectrum β-lactamases. Clinical Microbiology and Infection. 2008;**14**:75-81

[78] Poirel L, Menuteau O, Agoli N, Cattoen C, Nordmann P. Outbreak of extended-spectrum β-lactamase VEB-1-producing isolates of *Acinetobacter baumannii* in a French hospital. Journal of Clinical Microbiology. 2003;**41**(8):3542-3547

[79] Carattoli A. Resistance plasmid families in *Enterobacteriaceae*. Antimicrobial Agents and Chemotherapy. 2009;**53**(6):2227-2238

[80] Peirano G, Mulvey GL, Armstrong GD, Pitout JD. Virulence potential and adherence properties of *Escherichia coli* that produce CTX-M and NDM β-lactamases. Journal of Medical Microbiology. 2013;**62**(4):525-530

[81] Xia J, Sun J, Li L, Fang LX, Deng H, Yang RS, et al. First report of the IncI1/ST898 conjugative plasmid carrying rmtE2 16S rRNA methyltransferase gene in *Escherichia coli*. Antimicrobial Agents and Chemotherapy. 2015;**59**(12):7921-7922

[82] Aghapour Z, Gholizadeh P, Ganbarov K, Bialvaei AZ, Mahmood SS, Tanomand A, et al. Molecular mechanisms related to colistin resistance in Enterobacteriaceae. Infection and Drug Resistance. 2019;**12**:965

[83] Sköld O. Resistance to trimethoprim and sulfonamides. Veterinary Research. 2001;**32**(3-4):261-273

[84] Liu J, Keelan P, Bennett PM, Enne VI. Characterization of a novel macrolide efflux gene, mef (B), found linked to sul3 in porcine *Escherichia coli*. Journal of Antimicrobial Chemotherapy. 2009;**63**(3):423-426

[85] Boyd DA, Tyler S, Christianson S, McGeer A, Muller MP, Willey BM, et al. Complete nucleotide sequence of a 92-kilobase plasmid harboring the CTX-M-15 extended-spectrum beta-lactamase involved in an outbreak in long-term-care facilities in Toronto. Canada. Antimicrobial Agents and Chemotherapy. 2004;**48**(10):3758-3764

[86] Van Hoek AH, Mevius D, Guerra B, Mullany P, Roberts AP, Aarts HJ. Acquired antibiotic resistance genes: An overview. Frontiers in Microbiology. 2011;**2**:203

[87] Olaitan AO, Morand S, Rolain JM. Mechanisms of polymyxin resistance: Acquired and intrinsic resistance in bacteria. Frontiers in Microbiology. 2014;**5**:643

[88] Okusu H, Ma D, Nikaido H. AcrAB efflux pump plays a major role in the antibiotic resistance phenotype of *Escherichia coli* multiple-antibiotic-resistance (Mar) mutants. Journal of Bacteriology. 1996;**178**(1):306-308

[89] Karim A, Poirel L, Nagarajan S, Nordmann P. Plasmid-mediated extended-spectrum β-lactamase (CTX-M-3 like) from India and gene association with insertion sequence

IS Ecp1. FEMS Microbiology Letters. 2001;**201**(2):237-241

[90] Paterson DL, Bonomo RA. Extended-spectrum β-lactamases: A clinical update. Clinical Microbiology Reviews. 2005;**18**(4):657-686

[91] Lartigue MF, Poirel L, Aubert D, Nordmann P. In vitro analysis of IS Ecp1B-mediated mobilization of naturally occurring β-lactamase gene Bla CTX-M of Kluyvera ascorbata. Antimicrobial Agents and Chemotherapy. 2006;**50**(4):1282-1286

[92] Alton NK, Vapnek D. Transcription and translation of R-plasmid 538-1 DNA: Effects of mercury induction and analysis of polypeptides coded for by the r-determinant region. Plasmid. 1979;**2**(3):366-376

[93] Schwarz S, Kehrenberg C, Doublet B, Cloeckaert A. Molecular basis of bacterial resistance to chloramphenicol and florfenicol. FEMS Microbiology Reviews. 2004;**28**(5):519-542

[94] Butaye P, Devriese LA, Haesebrouck F. Antimicrobial growth promoters used in animal feed: Effects of less well-known antibiotics on gram-positive bacteria. Clinical Microbiology Reviews. 2003;**16**(2):175-188

[95] Perreten V, Boerlin P. A new sulphonamide resistance gene (sul3) in *Escherichia coli* is widespread in the pig population of Switzerland. Antimicrobial Agents and Chemotherapy. 2003;**47**(3):1169-1172

[96] Barbosa TM, Levy SB. Differential expression of over 60 chromosomal genes in *Escherichia coli* by constitutive expression of MarA. Journal of Bacteriology. 2000;**182**(12):3467-3474

[97] Cohen SP, McMurry LM, Levy SB. marA locus causes decreased expression of OmpF porin in multiple-antibiotic-resistant (mar) mutants of *Escherichia coli*. Journal of Bacteriology. 1988;**170**(12):5416-5422

[98] Fajardo A, Martinez-Martin N, Mercadillo M, Galán JC, Ghysels B, Matthijs S, et al. The neglected intrinsic resistome of bacterial pathogens. PLoS One. 2008;**3**(2):e1619

[99] Cox G, Wright GD. Intrinsic antibiotic resistance: Mechanisms, origins, challenges and solutions. International Journal of Medical Microbiology. 2013;**303**(6-7):287-292

[100] Ramsamy Y, Mlisana KP, Amoako DG, Abia AL, Ismail A, Allam M, et al. Mobile genetic elements-mediated Enterobacterales-associated carbapenemase antibiotic resistance genes propagation between the environment and humans: A One Health South African study. Science of the Total Environment. 2022;**806**:150641

[101] Poolman JT, Wacker M. Extraintestinal pathogenic Escherichia coli, a common human pathogen: challenges for vaccine development and progress in the field. The Journal of Infectious Diseases. 2016;**213**(1):6-13

[102] Knox JR. Extended-spectrum and inhibitor-resistant TEM-type beta-lactamases: Mutations, specificity, and three-dimensional structure. Antimicrobial Agents and Chemotherapy. 1995;**39**(12):2593-2601

[103] Martinez P, Garzón D, Mattar S. CTX-M-producing *Escherichia coli* and *Klebsiella pneumoniae* isolated from community-acquired urinary tract infections in Valledupar, Colombia. The Brazilian Journal of Infectious Diseases. 2012;**16**(5):420-425

[104] Saidani M, Messadi L, Mefteh J, Chaouechi A, Soudani A, Selmi R, et al.

Various Inc-type plasmids and lineages of *Escherichia coli* and *Klebsiella pneumoniae* spreading blaCTX-M-15, blaCTX-M-1 and mcr-1 genes in camels in Tunisia. Journal of Global Antimicrobial Resistance. 2019;**19**:280-283

[105] Novais Â, Cantón R, Coque TM, Moya A, Baquero F, Galán JC. Mutational events in cefotaximase extended-spectrum β-lactamases of the CTX-M-1 cluster involved in ceftazidime resistance. Antimicrobial Agents and Chemotherapy. 2008;**52**(7):2377-2382

[106] Delannoy S, Le Devendec L, Jouy E, Fach P, Drider D, Kempf I. Characterization of colistin-resistant *Escherichia coli* isolated from diseased pigs in France. Frontiers in Microbiology. 2017;**8**:2278

[107] Newton-Foot M, Snyman Y, Maloba MR, Whitelaw AC. Plasmid-mediated mcr-1 colistin resistance in Escherichia coli and Klebsiella spp. clinical isolates from the Western Cape region of South Africa. Antimicrobial Resistance & Infection Control. 2017;**6**(1):1-7

[108] Kolář M, Urbanek K, Látal T. Antibiotic selective pressure and development of bacterial resistance. International Journal of Antimicrobial Agents. 2001;**17**(5):357-363

[109] Balbin MM, Hull D, Guest C, Nichols L, Dunn R, Thakur S. Antimicrobial resistance and virulence factors profile of salmonella spp. and *Escherichia coli* isolated from different environments exposed to anthropogenic activity. Journal of Global Antimicrobial Resistance. 2020;**22**:578-583

[110] Shively NR, Buehrle DJ, Clancy CJ, Decker BK. Prevalence of inappropriate antibiotic prescribing in primary care clinics within a veterans affairs health care system. Antimicrobial Agents and Chemotherapy. 2018;**62**(8):e00337-e00318

[111] Tiseo K, Huber L, Gilbert M, Robinson TP, Van Boeckel TP. Global trends in antimicrobial use in food animals from 2017 to 2030. Antibiotics. 2020;**9**(12):918

[112] Browne AJ, Chipeta MG, Haines-Woodhouse G, Kumaran EP, Hamadani BH, Zaraa S, et al. Global antibiotic consumption and usage in humans, 2000-18: A spatial modelling study. The Lancet Planetary Health. 2021;**5**(12):e893-e904

[113] Schneiders T, Amyes SG, Levy SB. Role of AcrR and RamA in fluoroquinolone resistance in clinical *Klebsiella pneumoniae* isolates from Singapore. Antimicrobial Agents and Chemotherapy. 2003;**47**(9):2831-2837

[114] Wang H, Dzink-Fox JL, Chen M, Levy SB. Genetic characterization of highly fluoroquinolone-resistant clinical *Escherichia coli* strains from China: Role of acrR mutations. Antimicrobial Agents and Chemotherapy. 2001;**45**(5):1515-1521

[115] Piddock LJ. Mechanisms of fluoroquinolone resistance: An update 1994-1998. Drugs. 1999;**58**(2):11-18

[116] Lu J, Wang Y, Li J, Mao L, Nguyen SH, Duarte T, et al. Triclosan at environmentally relevant concentrations promotes horizontal transfer of multidrug resistance genes within and across bacterial genera. Environment International. 2018;**121**:1217-1226

[117] Dann AB, Hontela A. Triclosan: Environmental exposure, toxicity and mechanisms of action. Journal of Applied Toxicology. 2011;**31**(4):285-311

[118] Collignon P, Beggs JJ, Walsh TR, Gandra S, Laxminarayan R.

Anthropological and socioeconomic factors contributing to global antimicrobial resistance: A univariate and multivariable analysis. The Lancet Planetary Health. 2018;**2**(9):e398-e405

[119] Hu Y, Yang X, Lu N, Zhu B. The abundance of antibiotic resistance genes in human guts has correlation to the consumption of antibiotics in animal. Gut Microbes. 2014;**5**(2):245-249

[120] Clements A, Halton K, Graves N, Pettitt A, Morton A, Looke D, et al. Overcrowding and understaffing in modern health-care systems: Key determinants in meticillin-resistant *Staphylococcus aureus* transmission. The Lancet Infectious Diseases. 2008;**8**(7):427-434

[121] Kumarasamy KK, Toleman MA, Walsh TR, Bagaria J, Butt F, Balakrishnan R, et al. Emergence of a new antibiotic resistance mechanism in India, Pakistan, and the UK: A molecular, biological, and epidemiological study. The Lancet Infectious Diseases. 2010;**10**(9):597-602

[122] Singh BR. ESKAPE pathogens in animals and their antimicrobial drug resistance pattern. Journal of Dairy, Veterinary & Animal Research. 2018;7(3):1-0

[123] Lagier JC, Armougom F, Million M, Hugon P, Pagnier I, Robert C, et al. Microbial culturomics: Paradigm shift in the human gut microbiome study. Clinical Microbiology and Infection. 2012;**18**(12):1185-1193

[124] Berendonk TU, Manaia CM, Merlin C, Fatta-Kassinos D, Cytryn E, Walsh F, et al. Tackling antibiotic resistance: The environmental framework. Nature Reviews Microbiology. 2015;**13**(5):310-317

[125] Naghavi M, Murray CJ, Ikuta KS, Mestrovic T, Swetschinski L, Sartorius B. Global burden of antimicrobial resistance: Essential pieces of a global puzzle–Authors' reply. The Lancet. 2022;**399**(10344):2349-2350

Chapter 13

Phylotypes and Pathotypes of Diarrheagenic *Escherichia coli* of Gastroenteritis

Hadi Sajid Abdulabbas, Noor Al-Khafaji, Suhad Y. Abed, Hussein Al-Dahmoshi and Huda Najh Al-Baroody

Abstract

Escherichia coli responsible for wide range of common bacterial infections, the frequent one is gastroenteritis. Bacterial gastroenteritis mainly attributed to diarrheagenic *E. coli* and accompanied by diarrhea and vomiting. Actually pathogenic *E. coli* can be classified according to the site of infection whether it be within intestine (called intestinal *pathogenic E. coli* InPEC) or cause infection outside intestine (called extraintestinal pathogenic *E. coli* ExPEC). They are assigned to 4 main phylogenetic groups: InPEC include A and B1 while ExPEC have B2 and D groups. Seven Pathotypes have been assigned: Enteropathogenic *E. coli* (EPEC), enterotoxigenic *E. coli* (ETEC), enterohemorrhagic (Shiga-toxin producing *E. coli* (EHEC/STEC), enteroaggregative *E. coli* (EAEC), diffusely adherent *E. coli* (DAEC), enteroinvasive *E. coli* (EIEC) and adherent invasive *E. coli* (AIEC). The patho-phylotyping of diarrheagenic *E. coli* interaction along with antibiotic resistance and biofilm formation capacity may be valuable insight to know real threat of this pathogen and this is tried to be covered with this chapter. The results revealed that the among DEC, EPEC and ETEC were assigned in high rate to B1 followed by A, B2, D, E, C and F while EAEC show different assignment: D followed by B2, A, B1, C, E and F. The other DEC pathotypes showed different styles.

Keywords: Pathotypes, phylotypes, *E. coli*, antibiotic resistance, biofilm, InPEC, ExPEC

1. Introduction

Escherichia coli is commensal enterobacteial member emerging as opportunistic pathogen causing numerous of intestinal and extraintestinal infections. Concern Pathotypes of *E. coli*, they divided in to two main groups: Extraintestinal pathogenic *E. coli* (ExPEC) and intestinal pathogenic *E. coli* (InPEC). ExPEC cause range of infections for Human being outside intestine like meningitides as in neonatal meningitis *E. coli* (NMEC), urinary tract infection as in uropathogenic *E. coli* (UPEC) and sepsis-associated *E. coli* (SEPEC) [1–3]. InPEC includes 8 Pathotypes of diarrheagenic *E. coli* (DEC) including: enteropathogenic *E. coli* (EPEC), enterotoxigenic

arpA	*chuA*	*yjaA*	*TspE4.C2*	*arpAgpE*	*trpAgpC*	**Phylogroups and Clade**
+	—	—	—	—	—	A (subgroup A0)
+	—	+	—	—	—	A (subgroup A1)
+	—	—	+	—	—	B1
—	+	+	—	—	—	B2(subgroup $B2_2$)
—	+	+	+	—	—	B2(subgroup $B2_3$)
+	—	+	—	—	+	C
+	+	—	—	—	—	D (subgroup D1)
+	+	—	+	—	—	D (subgroup D2)
+	+	—	+	+	—	E
+	+	—	+	+	—	E
—	+	—	—	—	—	F

Table 1.
Genotypic profile for 6 phylogenetic groups and 5 clades of E. coli.

E. coli (ETEC), enteroaggregative *E. coli* (EAEC), diffusely adherent *E. coli* (DAEC), enteroinvasive *E. coli* (EIEC), enterohemorrhagic *E. coli* (EHEC) (also known as Shiga-toxin-producing *E. coli* (STEC) or, verotoxin-producing *E. coli* (VTEC), necrotoxic *E. coli* (NTEC) and adherent-invasive *E. coli* (AIEC) [4–6]. Two systems named as Clermont triplex and quadruplex PCR were used to assign the *E. coli* isolates to seven phylogroups (A, B1, B2, C, D, E and F) and five clades (clade I, II, III, IV and V) according to presence/absence of six genes: chuA, yjaA, TspE4.C2, arpA, *arpAgpEh and trpA* (trpAgpC) as mentioned in **Table 1** [7–9].

2. Diarrheagenic *E. coli* (DEC) phylotypes

Diarrheagenic *E. coli* (DEC) is a term referring to 8 or 9 pathotypes of *E. coli* causing diarrhea in travelers. It accounting for up to 40% of infantile and children diarrhea worldwide [10] and responsible for most of diarrheal outbreaks globally [11]. All DEC Pathotypes categorized according to molecular contents of some virulence genes: *eae* and *bfpA* for EPEC [12]; *st1a*, *st1b* and *ltb1* for ETEC; *aggR* for EAEC; *stx1* and *stx2* for EHEC/STEC; *ipaH* for EIEC; *daaD* for DAEC [13]. Many studies revealed that, the most prevalent diarrheagenic *E. coli* were ETEC, EPEC, EAEC, AIEC, EHEC/STEC and DAEC [14–17]. Concern phylogroups of EPEC, the results gathered from 18 studies for 433 EPEC isolates revealed that: B1 compile (39.3%), A(23.3%), B2(18.2%), D(8.5%), (6.2%, 3.5% and 0.9%) for E, C and F phylogroups respectively (**Table 2**) [18–35]. ETEC phylogroups includes: B1(43.3%), A(28.2%), D(14%), B2(13.4%), C(0.6%), E(0.6%) and F(0.0%) (**Table 3**) [18, 23–25, 27, 28, 30, 31, 33, 35–37]. Concern EAEC phylogroups the results showed that, among 399 EAEC isolates they assigned for D(32.8%), B2(30.5%)A(19.2), B1(16.2%), C(0.8%), E(0.5%) and F(0.0%) (**Table 4**) [18, 24, 25, 27, 28, 30–32, 35, 37–40]. **Table 5** showed the phylogroups of EIEC, EHEC/STEC, AIEC and DAEC pathotypes. EIEC phylogroups were A, B1, E, D, B2 and C compile (37.6%, 21.2%, 17.7%, 12.9%, 8.2% and 2.4% respectively). EHEC/STEC phylogroups were C, B1, B2, A, E, D and Fcompile (21.5%, 20.1, 18.7%, 16.2%, 13.4%, 7.7% and 2.4% respectively). AIEC phylogroups were B2, D, A and B1 compile (51.3%, 25.2%, 18.3% and 5.2%

Total isolate no.	ExPEC group		InPEC group					Country	Ref.
	B2	D	A	B1	C	E	F		
5	1	0	1	1	1	1	0	South Africa	[18]
53	2	2	9	38	0	0	2	Mexico	[19]
56	25	4	11	12	2	1	1	India	[20]
51	9	2	2	13	8	17	0	Iran	[21]
42	2	6	2	32	0	0	0	Iran	[22]
26	1	4	8	13	0	0	0	Iran	[23]
10	6	0	3	1	0	0	0	Iran	[24]
3	1	0	2	0	0	0	0	Iran	[25]
4	0	0	0	4	0	0	0	Iran	[26]
9	7	0	2	0	0	0	0	Iran	[27]
4	0	0	0	4	0	0	0	Bangladesh	[28]
31	3	0	6	17	0	5	0	Brazil	[29]
25	1	0	18	6	0	0	0	Brazil	[30]
21	3	3	12	3	0	0	0	Brazil	[31]
30	12	5	13	0	0	0	0	Egypt	[32]
32	6	10	6	2	4	3	1	Egypt	[33]
26	0	1	5	20	0	0	0	Italy	[34]
5	0	0	1	4	0	0	0	Switzerland	[35]
Total(433)	79	37	101	170	15	27	4		
%	18.2	8.5	23.3	39.3	3.5	6.2	0.9		

Table 2.
ExPEC and InPEC phylogroups of EPEC isolates.

Total isolate no.	ExPEC group		InPEC group					Country	Ref.
	B2	D	A	B1	C	E	F		
18	4	1	1	8	2	2	0	South Africa	[18]
32	7	7	11	7	0	0	0	Iran	[36]
13	1	1	10	1	0	0	0	Iran	[23]
6	3	0	3	0	0	0	0	Iran	[24]
83	12	16	34	21	0	0	0	Iran	[25]
52	7	7	22	16	0	0	0	Iran	[27]
94	3	9	0	82	0	0	0	Bangladesh	[28]
9	0	4	3	2	0	0	0	Brazil	[30]
19	5	1	9	4	0	0	0	Brazil	[31]
9	3	2	0	4	0	0	0	Egypt	[33]
5	0	0	1	4	0	0	0	Switzerland	[35]
4	1	0	3	0	0	0	0	Mexico	[37]
Total(344)	46	48	97	149	2	2	0		
%	13.4	14	28.2	43.3	0.6	0.6	0		

Table 3.
ExPEC and InPEC phylogroups of ETEC isolates.

Total isolate no.	ExPEC group		InPEC group					Country	Ref.
	B2	D	A	B1	C	E	F		
31	9	4	3	10	3	2	0	South Africa	[18]
35	16	8	6	6	0	0	0	Iran	[24]
27	3	6	11	7	0	0	0	Iran	[25]
23	7	14	2	0	0	0	0	Iran	[27]
1	0	0	0	1	0	0	0	Bangladesh	[28]
14	3	0	6	5	0	0	0	Brazil	[30]
31	17	1	13	0	0	0	0	Egypt	[32]
27	0	9	17	1	0	0	0	Brazil	[31]
54	31	19	4	0	0	0	0	Brazil	[38]
8	0	4	3	1	0	0	0	Switzerland	[35]
1	1	0	0	0	0	0	0	Mexico	[37]
9	3	6	0	0	0	0	0	Tunisia	[39]
138	32	60	12	34	0	0	0	India	[40]
Total(399)	122	131	77	65	3	2	0		
%	30.5	32.8	19.2	16.2	0.8	0.5	0		

Table 4.
ExPEC and InPEC phylogroups of EAEC isolates.

EIEC									
Total isolate no.	**ExPEC group**		**InPEC group**					**Country**	**Ref.**
	B2	**D**	**A**	**B1**	**C**	**E**	**F**		
6	0	3	1	2	0	0	0	Iran	[36]
51	1	0	25	9	1	15	0	Brazil	[41]
6	1	1	3	1	0	0	0	Iran	[24]
13	3	4	1	5	0	0	0	Iran	[25]
2	0	2	0	0	0	0	0	Iran	[27]
7	2	1	2	1	1	0	0	Egypt	[33]
Total (42)	7	11	32	18	2	15	0		
%	8.2	12.9	37.6	21.2	2.4	17.7	0		
EHEC/STEC									
5	3	0	0	0	1	1	0	South Africa	[18]
21	2	4	0	15	0	0	0	Iran	[26]
10	1	1	0	8	0	0	0	Bangladesh	[28]
22	5	4	2	3	7	0	1	Egypt	[33]
6	0	0	6	0	0	0	0	Mexico	[37]
5	2	3	0	0	0	0	0	Tunisia	[39]
135	26	4	21	16	37	27	4	Iran	[43]
5	0	0	5	0	0	0	0	Brazil	[44]
Total(209)	39	16	34	42	45	28	5		

EIEC									
Total isolate no.	**ExPEC group**		**InPEC group**					**Country**	**Ref.**
	B2	**D**	**A**	**B1**	**C**	**E**	**F**		
%	18.7	7.7	16.2	20.1	21.5	13.4	2.4		
AIEC									
26	14	10	2	0	0	0	0	Spain	[45]
30	22	4	4	0	0	0	0	Spain	[46]
59	23	15	15	6	0	0	0	Korea	[47]
Total(115)	59	29	21	6	0	0	0		
%	51.3	25.2	18.3	5.2	0	0	0		
DAEC									
11	3	0	5	3	0	0	0	Iran	[25]

Table 5.
ExPEC and InPEC phylogroups of EIEC, EHEC/STEC AIEC and DAEC isolates.

respectively) while DAEC phylogroups were A (5/11), B2(3/11) and B1(3/11) [18, 24–28, 33, 36, 37, 39–41, 43–47].

3. Uropathogenic *E. coli* (UPEC) phylotypes

Uropathogenic *E. coli* (UPEC) is the most common ExPEC causing UTI, one of the most frequent bacterial infections with significant morbidity worldwide, which

Total isolate no.	**ExPEC group**		**InPEC group**					**Ref.**
	B2	**D**	**A**	**B1**	**C**	**E**	**F**	
57	38	3	6	7	1	0	2	[51]
88	46	7	11	0	13	11	0	[52]
50	18	14	4	14	0	0	0	[53]
42	18	15	8	1	0	0	0	[54]
50	18	21	9	0	0	0	0	[55]
395	172	113	96	14	0	0	0	[56]
103	38	20	27	9	5	0	4	[57]
100	91	6	1	2	0	0	0	[58]
56	43	4	4	5	0	0	0	[59]
72	26	16	20	2	4	0	4	[60]
105	57	22	22	4	0	0	0	[61]
Total(1119)	566	241	208	60	23	11	10	
%	50.6	21.5	18.6	5.4	2.1	1	0.8	

Table 6.
Distribution of phylogenetic groups among UPEC isolates in Iraq.

may convey a sets of virulence factors and belonging to diverse phylogenetic groups [48]. UPEC as ExPEC must be belong to either B2 or D phylogroups. The B2 include 2 phylosubgroups: $B2_2$ (*chuA+, yjaA+, TspE4.C2-, arpA-, arpAgpE-, TrpAgpC-*) and $B2_3$ (*chuA+, yjaA+, TspE4.C2+, arpA-, arpAgpE-, TrpAgpC-*) while D have 2 phylosubgroups: D1 (*chuA+, yjaA-, TspE4.C2-, arpA+, arpAgpE-, TrpAgpC-*) and D2 (*chuA+, yjaA-, TspE4.C2-, arpA+, arpAgpE-, TrpAgpC+*) [49, 50].

Many Iraqi studies studied the assignment of UPEC to phylogroups. Since 2017–2022 only 11 Iraqi studies with valuable no. of UPEC isolates analyze 1119 isolates and assigning them to 7 phylogroups and they are found that: 566(50.6%) for B2, 241(21.5%) for D, 208(18.6%) for A, 60(5.4%) for B1, 23(2.1%) for C, 11(1%) for E and 10(0.8%) for F (**Table 6**) [51–61]. Concern Iranian studies, 19 study were conducted including 2313 UPEC isolates assigned to phylogroups as follow: 1190(51.4%) for B2, 481(20.8%) for D, 399(17.3%) for A, 190(8.2%) for B1, 26(1.1%) for C, 23(1%) for E and 4(0.2%) for F (**Table 7**) [25, 62–79]. Ten studies from Egypt deal with 892 UPEC isolates stated that, B2 compile (38.1%), D (15.2%), A (23.2%), B1 (14.3%), C (3.4%), E (2.8%) and F (2.9%) as mentioned in **Table 8** [25, 42, 49, 54, 76, 80–84]. Nine Turkish studies showed that, among 901 isolated of UPEC, (39.3%) were B2,

Total isolate no.	ExPEC group		InPEC group					Ref.
	B2	D	A	B1	C	E	F	
93	55	4	1	7	9	13	4	[62]
100	52	28	14	6	0	0	0	[63]
232	135	42	37	18	0	0	0	[64]
113	50	35	23	5	0	0	0	[65]
251	92	79	40	40	0	0	0	[66]
77	55	22	0	0	0	0	0	[67]
90	47	12	4	0	17	10	0	[68]
160	75	48	25	12	0	0	0	[69]
50	24	10	15	1	0	0	0	[70]
104	26	20	32	26	0	0	0	[25]
100	55	22	17	6	0	0	0	[71]
60	50	6	3	1	0	0	0	[72]
261	144	28	61	28	0	0	0	[73]
93	26	29	29	9	0	0	0	[74]
144	106	23	7	8	0	0	0	[75]
90	25	11	46	8	0	0	0	[76]
100	55	22	17	6	0	0	0	[77]
90	53	18	13	6	0	0	0	[78]
105	65	22	15	3	0	0	0	[79]
Total(2313)	1190	481	399	190	26	23	4	
%	51.4	20.8	17.3	8.2	1.1	1	0.2	

Table 7.
Distribution of phylogenetic groups among UPEC isolates in Iran.

Total isolate no.	ExPEC group		InPEC group					Ref.
	B2	D	A	B1	C	E	F	
175	113	33	19	10	0	0	0	[80]
42	18	15	8	1	0	0	0	[54]
59	42	0	0	0	0	0	17	[49]
103	24	17	56	6	0	0	0	[81]
22	7	9	4	2	0	0	0	[82]
144	40	12	15	38	15	17	7	[42]
105	32	17	26	30	0	0	0	[83]
104	26	20	32	26	0	0	0	[25]
90	25	11	46	8	0	0	0	[76]
48	13	2	1	7	15	8	2	[84]
Total(892)	340	163	207	128	30	25	26	
%	38.1	15.2	23.2	14.3	3.4	2.8	2.9	

Table 8.
Distribution of phylogenetic groups among UPEC isolates in Egypt.

Total isolate no.	ExPEC group		InPEC group					Ref.
	B2	D	A	B1	C	E	F	
97	20	37	29	11	0	0	0	[85]
50	22	13	11	4	0	0	0	[86]
153	58	48	41	6	0	0	0	[87]
45	19	4	1	0	1	19	1	[88]
146	52	43	36	15	0	0	0	[89]
69	16	31	15	7	0	0	0	[90]
43	18	2	23	0	0	0	0	[91]
156	95	48	8	5	0	0	0	[92]
142	54	58	20	10	0	0	0	[93]
Total(901)	354	284	184	58	1	19	1	
%	39.3	31.5	2.4	6.4	0.1	2.1	0.1	

Table 9.
Distribution of phylogenetic groups among UPEC isolates in Turkey.

(31.5%) for D, (2.4%) for A, (6.4%) for B1, (0.1%) for C, (2.1%) for E and (0.1%) for F phylogroups (**Table 9**) [85–93].

4. Conclusion

Data analysis of current study revealed that the among DEC, EPEC and ETEC were assigned in high rate to B1 followed by A, B2, D, E, C and F while EAEC show different assignment: D followed by B2, A, B1, C, E and F. The other DEC pathotypes showed different styles.

Conflict of interest

There is no 'conflict of interest' for this work.

Author details

Hadi Sajid Abdulabbas[1], Noor Al-Khafaji[2], Suhad Y. Abed[3], Hussein Al-Dahmoshi[2*] and Huda Najh Al-Baroody[4]

1 Faculty of Dentistry, University of Al-Ameed, Karbala, Iraq

2 Biology Department, College of Science, University of Babylon, Hilla, Iraq

3 Department of Biology, College of Science, Mustansiriyah University, Baghdad, Iraq

4 Department of Microbiology, College of Medicine, University of Al-Ameed, Karbala, Iraq

*Address all correspondence to: dr.dahmoshi83@gmail.com

References

[1] Denamur E, Clermont O, Bonacorsi S, Gordon D. The population genetics of pathogenic *Escherichia coli*. Nature Reviews. Microbiology. 2021;**19**(1):37-54. DOI: 10.1038/s41579-020-0416-x

[2] Connolly JPR, Turner NCA, Hallam JC, Rimbi PT, Flett T, McCormack MJ, et al. d-serine induces distinct transcriptomes in diverse Escherichia coli pathotypes. Microbiology (Reading). 2021;**167**(10):1-13. DOI: 10.1099/mic.0.001093

[3] Becker S, Saidenberg A, AHM v V, Stegger M, Johannesen TB, Semmler T, et al. Genomic analysis of the zoonotic ST73 lineage containing avian and human extraintestinal pathogenic *Escherichia coli* (ExPEC). Veterinary Microbiology. 2022;**267**:109372. DOI: 10.1016/j.vetmic.2022.109372

[4] Martinez-Medina M. Special issue: Pathogenic Escherichia coli: Infections and therapies. Antibiotics. 2021;**10**(2):112. DOI: 10.3390/antibiotics10020112

[5] Angulo-Zamudio UA, Gutiérrez-Jiménez J, Monroy-Higuera L, Flores-Villaseñor H, Leon-Sicairos N, Velazquez-Roman J, et al. Non-diarrheagenic and diarrheagenic E. coli carrying supplementary virulence genes (SVG) are associated with diarrhea in children from Mexico. Microbial Pathogenesis. 2021;**157**:104994. DOI: 10.1016/j.micpath.2021.104994

[6] Suleiman K, Kolo I, Mohammed SSD, Magaji YG. Bacterial diarrhea among infants in developing countries: An overview of diarrheagenic Escherichia coli (DEC). Gadau Journal of Pure and Allied Science. 2022;**1**(1):73-81. DOI: 10.54117/gjpas.v1i1.10

[7] Clermont O, Christenson JK, Denamur E, Gordon DM. The Clermont Escherichia coli phylo-typing method revisited: Improvement of specificity and detection of new phylo-groups. Environmental Microbiology Reports. 2013;**5**(1):58-65. DOI: 10.1111/1758-2229.12019

[8] Clermont O, Gordon D, Denamur E. Guide to the various phylogenetic classification schemes for Escherichia coli and the correspondence among schemes. Microbiology (Reading). 2015;**161**(5):980-988. DOI: 10.1099/mic.0.000063

[9] Bujňáková D, Karahutová L, Kmeť V. Escherichia coli specific virulence-gene markers analysis for quality control of ovine cheese in Slovakia. Microorganisms. 2021;**9**(9):1808. DOI: 10.3390/microorganisms9091808

[10] Miliwebsky E, Schelotto F, Varela G, Luz D, Chinen I, Piazza RM. Human diarrheal infections: Diagnosis of diarrheagenic *Escherichia coli* pathotypes. In: *Escherichia coli* in the Americas. Berlin: Springer; 2016. pp. 343-369

[11] Croxen MA, Law RJ, Scholz R, Keeney KM, Wlodarska M, Finlay BB. Recent advances in understanding enteric pathogenic Escherichia coli. Clinical Microbiology Reviews. 2013;**26**(4):822-880. DOI: 10.1128/CMR.00022-13

[12] Kotloff KL, Nataro JP, Blackwelder WC, Nasrin D, Farag TH, Panchalingam S, et al. Burden and aetiology of diarrhoeal disease in infants and young children in developing countries (the global enteric multicenter study, GEMS): A prospective, case-control study. Lancet. 2013;**382**(9888):209-222. DOI: 10.1016/S0140-6736(13)60844-2

[13] Guion CE, Ochoa TJ, Walker CM, Barletta F, Cleary TG. Detection of diarrheagenic Escherichia coli by use of melting-curve analysis and real-time multiplex PCR. Journal of Clinical Microbiology. 2008;**46**(5):1752-1757

[14] Asadi Z, Ghanbarpour R, Kalantar-Neyestanaki D, Alizade H. Determination of extended-spectrum β-lactamase producing and hybrid pathotypes of Escherichia coli isolates from diarrheic samples. Gene Reports. 2022;**27**:101583. DOI: 10.1016/j.genrep.2022.101583

[15] Moyo SJ, Maselle SY, Matee MI, Langeland N, Mylvaganam H. Identification of diarrheagenic Escherichia coli isolated from infants and children in Dar es Salaam, Tanzania. BMC Infectious Diseases. 2007;**7**(1):92. DOI: 10.1186/1471-2334-7-92

[16] Franzolin MR, Alves RC, Keller R, Gomes TA, Beutin L, Barreto ML, et al. Prevalence of diarrheagenic Escherichia coli in children with diarrhea in Salvador, Bahia, Brazil. Memórias do Instituto Oswaldo Cruz. 2005;**100**(4):359-363. DOI: 10.1590/s0074-02762005000400004

[17] Abdlla YA, Al-Sanjary RA. The molecular identification of diarrheagenic Escherichia coli (DEC) isolated from meat and meat products. Iraqi Journal of Veterinary Sciences. 2022;**37**(1):9-15. DOI: 10.33899/ijvs.2022.133244.2192

[18] Alfinete NW, Bolukaoto JY, Heine L, Potgieter N, Barnard TG. Virulence and phylogenetic analysis of enteric pathogenic Escherichia coli isolated from children with diarrhoea in South Africa. International Journal of Infectious Diseases. 2022;**114**:226-232. DOI: 10.1016/j.ijid.2021.11.017

[19] Otero-García A, Pérez-Vargas AK, Eslava-Campos C, Licona-Moreno D, León LA, Pérez-Soto G, et al. Molecular genotyping of Escherichia coli o111 isolated from children with diarrhea in rural and urban areas. Europub Journal of Health Research. 2022;**3**(3):421-435. DOI: 10.54747/ejhrv3n3-009

[20] Singh T, Das S, Ramachandran VG, Wani S, Shah D, Maroof KA, et al. Distribution of Integrons and phylogenetic groups among Enteropathogenic Escherichia coli isolates from children <5 years of age in Delhi, India. Frontiers in Microbiology. 2017;**8**:561. DOI: 10.3389/fmicb.2017.00561

[21] Jafari E, Mostaan S, Bouzari S. Characterization of antimicrobial susceptibility, extended-spectrum β-lactamase genes and phylogenetic groups of enteropathogenic Escherichia coli isolated from patients with diarrhea. Osong Public Health and Research Perspectives. 2020;**11**(5):327-333. DOI: 10.24171/j.phrp.2020.11.5.09

[22] Memariani M, Najar-Peerayeh S, Zahraei Salehi TZ. Multi locus vntR (MLVA) typing and detection of the OI-122 pathogenicity island in typical and atypical enteropathogenic Escherichia coli isolated from children with acute diarrhea. Archives of Clinical Infectious Diseases. 2019;**14**(3):1-7. DOI: 10.5812/archcid.65855

[23] Taghadosi R, Shakibaie MR, Hosseini-Nave H. Antibiotic resistance, ESBL genes, integrons, phylogenetic groups and MLVA profiles of Escherichia coli pathotypes isolated from patients with diarrhea and farm animals in south-east of Iran. Comparative Immunology, Microbiology and Infectious Diseases. 2019;**63**:117-126. DOI: 10.1016/j.cimid.2019.01.004

[24] Mahdavi Broujerdi S, Roayaei Ardakani M, Rezatofighi SE. Characterization of diarrheagenic

Escherichia coli strains associated with diarrhea in children, Khouzestan, Iran. Journal of Infectious Development Ctries. 2018;**12**(8):649-656. DOI: 10.3855/jidc.9538

[25] Salmani H, Azarnezhad A, Fayazi MR, Hosseini A. Pathotypic and phylogenetic study of diarrheagenic Escherichia coli and uropathogenic E. coli using multiplex polymerase chain reaction. Jundishapur. Journal of Microbiology. 2016;**9**(2):e28331. DOI: 10.5812/jjm.28331

[26] Staji H, Badagliacca P, Salehi TZ, Lopes F, Iorio M, Tonelli A, et al. Pathotyping of diarrhoeagenic cattle Escherichia coli strains isolated in the province of Tehran, Iran. Veterinaria Italiana. 2017;**53**(4):345-356

[27] Alizade H, Ghanbarpour R, Aflatoonian MR. Molecular study on diarrheagenic Escherichia coli pathotypes isolated from under 5 years old children in southeast of Iran. Asian Pacific Journal of Tropical Disease. 2014;**4**:S813-S817. DOI: 10.1016/S2222-1808(14)60733-7

[28] Ferdous J, Rashid RB, Sultana R, Saima S, Jahan Prima M, Begum A, et al. Is it human or animal? The origin of pathogenic E. coli in the drinking water of a low-income Urban Community in Bangladesh. Trop med. Infectious Diseases. 2021;**6**(4):181. DOI: 10.3390/tropicalmed6040181

[29] Martins FH, Nepomuceno R, Piazza RMF, Elias WP. Phylogenetic distribution of tir-cytoskeleton coupling protein (tccP and tccP2) genes in atypical enteropathogenic Escherichia coli. FEMS Microbiology Letters. 2017;**364**(11):1-7. DOI: 10.1093/femsle/fnx101

[30] Coelho CVL, Gomes TAT, Vieira MAM, Rosa ACP, Marinho DP, Söhsten BLV, et al. Brazilian indigenous children as carriers of Diarrheagenic *Escherichia coli* Pathotypes. Advances in Infectious Disease. 2018;**08**(4):200-216. DOI: 10.4236/aid.2018.84017

[31] Regua-Mangia AH, Irino K, da Silva PR, Pimentel Bezerra RM, Santos Périssé AR, Teixeira LM. Molecular characterization of uropathogenic and diarrheagenic *Escherichia coli* pathotypes. Journal of Basic Microbiology. 2010;**50**(Suppl 1):S107-S115. DOI: 10.1002/jobm.200900364

[32] Khairy RMM, Fathy ZA, Mahrous DM, Mohamed ES, Abdelrahim SS. Prevalence, phylogeny, and antimicrobial resistance of Escherichia coli pathotypes isolated from children less than 5 years old with community acquired-diarrhea in upper Egypt. BMC Infectious Diseases. 2020;**20**(1):908. DOI: 10.1186/s12879-020-05664-6

[33] El-Shaer S, Abdel-Rhman SH, Barwa R, Hassan R. Virulence characteristics, serotyping and phylogenetic typing of clinical and environmental Escherichia coli isolates. Jundishapur Journal of Microbiology. 2018;**11**(12):1-12. DOI: 10.5812/jjm.82835

[34] Badagliacca P, Agnoletti F, Drigo I, Merildi V, Lopes F, Pompilii C, et al. Virulence gene profiles of rabbit enteropathogenic *Escherichia coli* strains isolated in northern Italy. Veterinaria Italiana. 2018;**54**(3):189-196. DOI: 10.12834/VetIt.859.4260.2

[35] Müller A, Stephan R, Nüesch-Inderbinen M. Distribution of virulence factors in ESBL-producing Escherichia coli isolated from the environment, livestock, food and humans. Science of the Total Environment. 2016;**541**:667-672. DOI: 10.1016/j.scitotenv.2015.09.135

[36] Hoseinzadeh T, Ghanbarpour R, Rokhbakhsh-Zamin F. Phylogenetic

background of enterotoxigenic and enteroinvasive Escherichia coli from patients with diarrhea in Sirjan, Iran. Iran Journal of Microbiology. 2016;**8**(3):187-192

[37] Bautista-Trujillo GU, Gutiérrez-Miceli FA, Mandujano-García L, Oliva-Llaven MA, Ibarra-Martínez C, Mendoza-Nazar P, et al. Captive green iguana carries diarrheagenic Escherichia coli pathotypes. Frontiers in Veterinary Science. 2020;7:99. DOI: 10.3389/fvets.2020.00099

[38] AMARAL MSMG. Genes relacionados à virulência, resistência aos antimicrobianos, filogrupos genéticos e resposta imunoinflamatória associadas com cepas de Escherichia coli enteroagregativa isoladas de crianças nutridas e desnutridas em Fortaleza-Ce, Brasil [dissertação] (Mestrado em Microbiologia Médica) - Faculdade de Medicina. Fortaleza: Universidade Federal do Ceará; 2018. p. 98 f

[39] Kilani H, Abbassi MS, Ferjani S, Ben Salem R, Mansouri R, Ben Chehida N, et al. Diverse Escherichia coli pathovars of phylogroups B2 and D isolated from animals in Tunisia. Journal of Infection in Developing Countries. 2017;**11**(7):549-556. DOI: 10.3855/jidc.8579

[40] Modgil V, Mahindroo J, Narayan C, Kalia M, Yousuf M, Shahi V, et al. Comparative analysis of virulence determinants, phylogroups, and antibiotic susceptibility patterns of typical versus atypical enteroaggregative E. coli in India. PLoS Neglected Tropical Diseases. 2020;**14**(11):e0008769. DOI: 10.1371/journal.pntd.0008769

[41] da Silva LC, de Mello Santos AC, Silva RM. Uropathogenic Escherichia coli pathogenicity islands and other ExPEC virulence genes may contribute to the genome variability of enteroinvasive E. coli. BMC Microbiology. 2017;**17**(1):68. DOI: 10.1186/s12866-017-0979-5

[42] El-Baz R, Said HS, Abdelmegeed ES, Barwa R. Characterization of virulence determinants and phylogenetic background of multiple and extensively drug resistant Escherichia coli isolated from different clinical sources in Egypt. Applied Microbiology and Biotechnology. 2022;**106**(3):1279-1298. DOI: 10.1007/s00253-021-11740-x

[43] Jafari E, Oloomi M, Bouzari S. Characterization of antimicrobial susceptibility, extended-spectrum β-lactamase genes and phylogenetic groups of shigatoxin producing Escherichia coli isolated from patients with diarrhea in Iran. Annals of Clinical Microbiology and Antimicrobials. 2021;**20**(1):24. DOI: 10.1186/s12941-021-00430-1

[44] Rebello RC, Regua-Mangia AH. Potential enterovirulence and antimicrobial resistance in Escherichia coli isolates from aquatic environments in Rio de Janeiro, Brazil. Science of the Total Environment. 2014;**490**:19-27. DOI: 10.1016/j.scitotenv.2014.04.040

[45] Céspedes S, Saitz W, Del Canto F, De la Fuente M, Quera R, Hermoso M, et al. Genetic diversity and virulence determinants of Escherichia coli strains isolated from patients with Crohn's disease in Spain and Chile. Frontiers in Microbiology. 2017;**8**:639. DOI: 10.3389/fmicb.2017.00639

[46] Camprubí-Font C, Lopez-Siles M, Ferrer-Guixeras M, Niubó-Carulla L, Abellà-Ametller C, Garcia-Gil LJ, et al. Comparative genomics reveals new single-nucleotide polymorphisms that can assist in identification of adherent-invasive Escherichia coli. Scientific Reports. 2018;**8**(1):2695. DOI: 10.1038/s41598-018-20843-x

[47] Lee JG, Han DS, Jo SV, Lee AR, Park CH, Eun CS, et al. Characteristics and pathogenic role of adherent-invasive Escherichia coli in inflammatory bowel disease: Potential impact on clinical outcomes. PLoS One. 2019;**14**(4):e0216165. DOI: 10.1371/journal.pone.0216165

[48] Rezatofighi SE, Mirzarazi M, Salehi M. Virulence genes and phylogenetic groups of uropathogenic Escherichia coli isolates from patients with urinary tract infection and uninfected control subjects: A case-control study. BMC Infectious Diseases. 2021;**21**(1):361. DOI: 10.1186/s12879-021-06036-4

[49] Hassuna NA, Khairalla AS, Farahat EM, Hammad AM, Abdel-Fattah M. Molecular characterization of extended-spectrum β lactamase- producing E. coli recovered from community-acquired urinary tract infections in upper Egypt. Scientific Reports. 2020;**10**(1):2772. DOI: 10.1038/s41598-020-59772-z

[50] Haghighatpanah M, Zeighami H, Mozaffari Nejad AS, Hajipour N. Phylogenetic groups and antimicrobial resistance characteristics of Escherichia coli strains isolated from clinical samples in North Iran. Arab J Gastroenterology. 2022;**23**(2):102-107. DOI: 10.1016/j.ajg.2022.02.003

[51] Al-Guranie DR, Al-Mayahie SM. Prevalence of E. coli ST131 among uropathogenic E. coli isolates from Iraqi patients in Wasit Province, Iraq. International Journal of Microbiology. 2020;**2020**:8840561. DOI: 10.1155/2020/8840561

[52] Allami M, Bahreini M, Sharifmoghadam MR. Antibiotic resistance, phylogenetic typing, and virulence genes profile analysis of uropathogenic Escherichia coli isolated from patients in southern Iraq. Journal of Applied Genetics. 2022;**63**(2):401-412. DOI: 10.1007/s13353-022-00683-2

[53] Ahmed NA, Mahmood SS, Abbas AH. A comparative study of some virulence factors and phylogenetic characterization of Escherichia coli isolates causing urinary tract infection and the commensal gut. Iraqi Journal of Agriculture Science. 2019;**50**(3):1193-1198

[54] Mahmoud AT, Salim MT, Ibrahem RA, Gabr A, Halby HM. Multiple drug resistance patterns in various phylogenetic groups of hospital-acquired uropathogenic E. coli isolated from cancer patients. Antibiotics (Basel). 2020;**9**(3):108. DOI: 10.3390/antibiotics9030108

[55] Hama-Soor TA, Salih SS, Muhamad SJ. The β-lactamase profile of Escherichia coli isolates from patients with urinary tract infections in teaching Hospital in Sulaimani, Iraq. Annals Parasitology. 2021;**67**(4):659-670. DOI: 10.17420/ap6704.382

[56] Pirko EY, Ali MRN. The relationship between phylogenic typing and antimicrobial susceptibility patterns for Escherichia coli isolated from UTIs atMany hospitals in Baghdad city. Nursing National Iraqi Specialties. 2017;**30**(2):1-2

[57] Gatya Al-Mayahie SM, Al-Guranie DRT, Hussein AA, Bachai ZA. Prevalence of common carbapenemase genes and multidrug resistance among uropathogenic Escherichia coli phylogroup B2 isolates from outpatients in Wasit Province/ Iraq. PLoS One. 2022;**17**(1):e0262984. DOI: 10.1371/journal.pone.0262984

[58] Hussein FA, Alwash MS. Molecular detection of extended spectrum

β-lactamase genes in Escherichia coli isolates from urinary tract infected patients. Journal of Applied National Sciences. 2022;**14**(4):1485-1492

[59] Al-Zuhairy AM, Al-Dahmoshi HO. CTX-M genotyping among cystitis associated cefotaxime-resistant uropathogenic Escherichia coli (CRUPEC). Journal of Pure and Applied Microbiology. 2018;**12**(3):1515-1523. DOI: 10.22207/JPAM.12.3.56

[60] Bachai ZA, Al-Mayahie SM. Molecular comparison of adhesins of uropathogenic Escherichia coli isolates from patients with first episode and recurrent urinary tract infections. Indian Journal of National Science. 2019;**9**(52):16529-16535

[61] Merza NS. Prevalence and molecular characterization of Fim H gene in Escherichia coli isolates recovered from patients with utis. Medicine Journal of Babylon. 2017;**14**(3):470-477

[62] Iranpour D, Hassanpour M, Ansari H, Tajbakhsh S, Khamisipour G, Najafi A. Phylogenetic groups of Escherichia coli strains from patients with urinary tract infection in Iran based on the new Clermont phylotyping method. BioMed Research International. 2015;**2015**:846219. DOI: 10.1155/2015/846219

[63] Salehzadeh A, Zamani H. Characterization of (uropathogenic) E. coli isolated from urinary tract infections: Phylogenetic typing and distribution of virulence-associated traits. British Journal of Biomedical Science. 2018;**75**(1):40-42. DOI: 10.1080/09674845.2017.1336834

[64] Farajzadah Sheikh AF, Goodarzi H, Yadyad MJ, Aslani S, Amin M, Jomehzadeh N, et al. Virulence-associated genes and drug susceptibility patterns of uropathogenic Escherichia coli isolated from patients with urinary tract infection. Infection and Drug Resistance. 2019;**12**:2039-2047. DOI: 10.2147/IDR.S199764

[65] Bakhtiari S, Mahmoudi H, Seftjani SK, Amirzargar MA, Ghiasvand S, Ghaffari ME, et al. Antibiotic resistance pattern and phylogenetic groups of the uropathogenic Escherichia coli isolates from urinary tract infections in Hamedan, west of Iran. Iran Journal of Microbiology. 2020;**12**(5):388-394. DOI: 10.18502/ijm.v12i5.4598

[66] Hashemizadeh Z, Kalantar-Neyestanaki D, Mansouri S. Association between virulence profile, biofilm formation and phylogenetic groups of Escherichia coli causing urinary tract infection and the commensal gut microbiota: A comparative analysis. Microbial Pathogenesis. 2017;**110**:540-545. DOI: 10.1016/j.micpath.2017.07.046

[67] Abdi HA, Rashki A. Comparison of virulence factors distribution in uropathogenic E. coli isolates from phylogenetic groups B2 and D. International. Journal of Enteric Pathogens. 2016;**2**(4):21725, 6

[68] Boroumand M, Naghmachi M, Ghatee MA. Detection of phylogenetic groups and drug resistance genes of Escherichia coli causing urinary tract infection in Southwest Iran. Jundishapur Journal of Microbiology. 2021;**14**(2):1-9. DOI: 10.5812/jjm.112547

[69] Staji H, Rassouli M, Jourablou S. Comparative virulotyping and phylogenomics of Escherichia coli isolates from urine samples of men and women suffering urinary tract infections. Iranian Journal of Basic Medical Sciences. 2019;**22**(2):211-214. DOI: 10.22038/ijbms.2018.28360.6880

[70] Javed S, Mirani ZA, Pirzada ZA. Phylogenetic group B2 expressed significant biofilm formation among drug resistant Uropathogenic Escherichia coli. Libyan Journal of Medicine. 2021;**16**(1):1845444. DOI: 10.1080/19932820.2020.1845444

[71] Rahdar M, Rashki A, Miri HR, Rashki Ghalehnoo MR. Detection of pap, sfa, afa, foc, and fim adhesin-encoding operons in uropathogenic Escherichia coli isolates collected from patients with urinary tract infection. Jundishapur Journal of Microbiology. 2015;**8**(8):e22647. DOI: 10.5812/jjm.22647

[72] Ranjbar R, Nazari S, Farahani O. Phylogenetic analysis and antimicrobial resistance profiles of Escherichia coli strains isolated from UTI-suspected patients. Iranian Journal of Public Health. 2020;**49**(9):1743-1749. DOI: 10.18502/ijph.v49i9.4094

[73] Norouzian H, Katouli M, Shahrokhi N, Sabeti S, Pooya M, Bouzari S. The relationship between phylogenetic groups and antibiotic susceptibility patterns of Escherichia coli strains isolated from feces and urine of patients with acute or recurrent urinary tract infection. Iran Journal of Microbiology. 2019;**11**(6):478-487. DOI: 10.18502/ijm.v11i6.2219

[74] Staji H, Khoshgoftar J, Javaheri Vayeghan A, Salimi Bejestani MR. Phylogenetic grouping and assessment of virulence genotypes, with antibiotic resistance patterns, of Escherichia coli strains implicated in female urinary tract infections. The International Journal of Enteric Pathogens. 2016;**4**(1):4. DOI: 10.17795/ijep31609

[75] Boroumand M, Sharifi A, Ghatei MA, Sadrinasab M. Evaluation of biofilm formation and virulence genes and association with antibiotic resistance patterns of uropathogenic Escherichia coli strains in southwestern Iran. Jundishapur Journal of Microbiology. 2021;**14**(9):1-10. DOI: 10.5812/jjm.117785

[76] El-Mahdy R, Mahmoud R, Shrief R. Characterization of E. coli phylogroups causing catheter-associated urinary tract infection. Infection and Drug Resistance. 2021;**14**:3183-3193. DOI: 10.2147/IDR.S325770

[77] Abdi HA, Rashki A. Relationship between phylogenetic group and distribution of virulence genes of Escherichia coli isolated from patients with urinary tract infection. Journal of Gorgan University of Medical Sciences. 2015;**17**(2):92-97

[78] Bahadori M, Motamedifar M, Derakhshandeh A, Firouzi R, Motamedi Boroojeni A, Alinejad M, et al. Genetic relatedness of the Escherichia coli fecal population and strains causing urinary tract infection in the same host. MicrobiologyOpen. 2019;**8**(6):e00759. DOI: 10.1002/mbo3.759

[79] Eghbalpour F, Vahdat S, Shahbazi R, Mohebi S, Kholdi S, Hadadi M, et al. Pathogenic features of urinary Escherichia coli strains causing asymptomatic bacteriuria during pregnancy. Gene Reports. 2022;**27**:101559. DOI: 10.1016/j.genrep.2022.101559

[80] Gawad WE, Helmy OM, Tawakkol WM, Hashem AM. Antimicrobial resistance, biofilm formation, and phylogenetic grouping of uropathogenic Escherichia coli isolates in Egypt: The role of efflux pump-mediated resistance. Jundishapur Journal of Microbiology. 2018;**11**(2):e14444. DOI: 10.5812/jjm.14444

[81] Khairy RM, Mohamed ES, Abdel Ghany HM, Abdelrahim SS. Phylogenic

classification and virulence genes profiles of uropathogenic E. coli and diarrhegenic E. coli strains isolated from community acquired infections. PLoS One. 2019;**14**(9):e0222441. DOI: 10.1371/journal.pone.0222441

[82] Tharwat N, El-Sherif R, Elnagdy S, Marzaban R, Amer S. Virulent Escherichia coli strains among Egyptian patients with acute diarrhoea versus urinary tract infection, and their antibiotic susceptibility. Arabian Journal of Gastroenterology. 2019;**20**(2):74-80. DOI: 10.1016/j.ajg.2019.01.003

[83] Abdelrahim SS, Fouad M, Abdallah N, Ahmed RF, Zaki S. Comparative study of CTX-M-15 producing Escherichia coli ST131 clone isolated from urinary tract infections and acute diarrhoea. Infection and Drug Resistance. 2021;**14**:4027-4038. DOI: 10.2147/IDR.S325669

[84] Ramadan HH, Jackson CR, Taha SA, Moawad AA, Barrett JB, Woodley TA. Contribution of healthy chickens to antimicrobial-resistant Escherichia coli associated with human extraintestinal infections in Egypt. Vector Borne and Zoonotic Diseases. 2018;**18**(8):408-416. DOI: 10.1089/vbz.2017.2237

[85] Bozcal E, Eldem V, Aydemir S, Skurnik M. The relationship between phylogenetic classification, virulence and antibiotic resistance of extraintestinal pathogenic Escherichia coli in İzmir province, Turkey. PeerJ. 2018;**6**:e5470. DOI: 10.7717/peerj.5470

[86] Oktay Gultekin E, Tezcan Ulger S, Delialioğlu N. Distribution of pathogenicity island markers and virulence factors genes of extraintestinal pathogenic Escherichia coli isolates. Jundishapur Journal of Microbiology. 2022;**15**(4):1-8. DOI: 10.5812/jjm-121044

[87] Yılmaz EŞ, Aslantaş Ö. Phylogenetic group/subgroups distributions, virulence factors, and antimicrobial susceptibility of *Escherichia coli* strains from urinary tract infections in Hatay. Revista da Sociedade Brasileira de Medicina Tropical. 2020;**53**:e20190429. DOI: 10.1590/0037-8682-0429-2019

[88] Gümüş D, Kalaycı Yüksek F, Camadan FD, Oral M, Macunluoğlu AC, Küçüker M. High prevalence of class I and class II integrons in uropathogenic *E. coli* strains (UPECS) and their relationship with antibiotic resistance, phylogeny and virulence. jmed. 2022;**85**(1):77-85. DOI: 10.26650/IUITFD.984487

[89] Er DK, Dundar D, Uzuner H, Osmani A. Relationship between phylogenetic groups, antibiotic resistance and patient characteristics in terms of adhesin genes in cystitis and pyelonephritis isolates of *Escherichia coli*. Microbial Pathogenesis. 2015;**89**:188-194. DOI: 10.1016/j.micpath.2015.10.014

[90] Giray B, Uçar FB, Aydemir SŞ. Genotypic analysis of *Escherichia coli* strains that cause urosepsis in the Aegean region. Turkish Journal of Medical Science. 2016;**46**(5):1518-1527. DOI: 10.3906/sag-1507-114

[91] Müştak HK, Günaydin E, Kaya İB, Salar MÖ, Babacan O, Önat K, et al. Phylo-typing of clinical *Escherichia coli* isolates originating from bovine mastitis and canine pyometra and urinary tract infection by means of quadruplex PCR. The Veterinary Quarterly. 2015;**35**(4):194-199. DOI: 10.1080/01652176.2015.1068963

[92] Onlen Guneri CO, Koksal F, Kizilyildirim S, Bedir B, Nagiyev T. The distribution of cytotoxic necrotizing factors (CNF-1, CNF-2, CNF-3) and cytolethal distending toxins (CDT-1,

CDT-2, CDT-3, CDT-4) in *Escherichia coli* isolates isolated from extraintestinal infections and the determination of their phylogenetic relationship by PFGE. International Journal of Clinical Practice. 2022;**2022**:7200635. DOI: 10.1155/2022/7200635

[93] Gundogdu A, Bolkvadze D, Kilic H. In vitro effectiveness of commercial bacteriophage cocktails on diverse extended-spectrum beta-lactamase producing *Escherichia coli* strains. Frontiers in Microbiology. 2016;7:1761. DOI: 10.3389/fmicb.2016.01761